Making Your Home Work For You

A DIY Smart Home Guide, Volume 1

Ronald Nutter

Published by TechBytes Press, 2018.

Publisher's Cataloging-In-Publication Data
(Prepared by The Donohue Group, Inc.)

Names: Nutter, Ronald, 1960- author.
Title: Making your home work for you / Ronald Nutter.
Description: Interactive edition. | [Lee's Summit, Missouri] : TechBytes Press, 2018. | Series: A DIY smart home guide ; volume 1 | The book contains URLs and QR codes that will link the reader to accompanying materials.
Identifiers: ISBN 9781947060180 (hardcover) | ISBN 9781947060081 (paperback) | ISBN 9781947060098 (ebook)
Subjects: LCSH: Home automation. | Home computer networks. | Household electronics. | Dwellings--Security measures.
Classification: LCC TK7881.25 .N88 2018 (print) | LCC TK7881.25 (ebook) | DDC 643.6--dc23

This book is a result of research that I have done and/or products that I am personally using. Product names and/or Company names associated with those products are as I found them. Any resemblance to actual persons, living or dead, or to actual events or locales is entirely coincidental.

TechBytes
Press
http://www.techbytespress.com

Version 2018.11.30

Disclaimer

Unfortunately in this world of lawsuits, I want to make the following disclaimer. Everything that I have written about here is about my own experiences or as a result of research that I did. Your results may vary from mine. What you will read about here are steps I have taken for my own emergency preparations. If I haven't ordered something but found that it was something that I would be looking at in the future, I will indicate that.

Disclosure of Material Connection: Some or all of the links in this book are "affiliate links." This means if you click on the link and purchase the item, I will receive an affiliate commission. This will not affect the price you pay for the item you purchased. Regardless, I only recommend products or services I use or have used personally and believe will add value to my readers. I am disclosing this in accordance with the Federal Trade Commission's 16 CFR, Part 255: "Guides Concerning the Use of Endorsements and Testimonials in Advertising."

Foreword

Having grown up in a the pre-Internet world, I can totally appreciate how overwhelming change can be. Today, change happens so fast that it's almost overwhelming to keep up with all the new technologies and digital options available at our fingertips. It's only been about 10 years since the iPhone was released. That one device truly revolutionized our ability to communicate. It's so much more than just the phone without a dial... it is literally a computer in our pocket, and it has continued to grow as an integral part and a hub for our lives.

At the risk of dating myself, I grew up in the world where every telephone was hardwired. I even lived in places where party lines were still a thing (were multiple people and families share the same phone connection, and you had to wait for others to finish their conversation speak for you use the phone). I also remember when my dad brought home the first calculator. It was a Texas instruments model that cost hundreds of dollars. Today you can buy one with 1000x the power and capabilities at the dollar store.

Some of you may even remember eight tracks, cassette tapes, and even CDs. These have all been replaced with devices that fit in our pocket or even on our wrist in the form of watch. I remember reading the Sunday new comics and seeing Dick Tracy communicate into his wristwatch. Now I marvel at the fact that I can actually take phone calls on my iWatch today.

The first computer I ever purchased was an Atari 400 where I learned to program in BASIC. Then I bought a Commodore 64 and then a 128, where I learned to make music and so much more. The first serious computer I got in the 1980s what's an IBM knockoff made by AT&T with Windows 3.1 (the first graphical user interface computer with a mouse that I ever used).

Today my watch, my phone, my computer, my TV, even some of my home appliances can now be controlled by the sound of my voice. These are the things that a little boy grew up dreaming about that I saw in the 60s while watching "Star Trek", "Lost In Space", and many other TV shows and movies about the future. Other than traveling light years in a metal box, we have exceeded a lot of what the shows predicted.

Embracing The Present and the Future

Although the ability to control a lightbulb with your voice may feel like a cool party trick to impress your friends, it actually has some incredible value to today's society. For those with limited mobility, it can be a gift from Heaven. The ability to control voice response technology removes some barriers to physical limitations. To the handicapped, it can be more than just a convenience, it can be a necessity.

With the advent of stronger and faster Wi-Fi connectivity within our homes and throughout the outside world, we have capabilities that we never could've imagined when that first iPhone was released. Even though wired signal connections provide more stability, reliability, and security, wireless technologies will quickly making hard wired connections less and less necessary. Having a Wi-Fi connection in your home, is not just becoming more commonplace but soon to become a necessity. Today, home security, garage doors, appliances, home entertainment, and yes lightbulbs can all be controlled to the sound of your voice on your watch or phone.

One barrier and/or complaint that people may have is that the cost of this Advanced Technology is too expensive for the average person. But I can assure you, that today is not the bleeding edge but the Renaissance. Not that long ago, DVD players where $1000. Now you can walk into a Walgreens or CVS and pick them up on sale for $30 dollars or less. The first 40 inch flat screen TVs were $25,000. Now you can find them for under $200 on Amazon and have them delivered to your home for free.

Even something as simple as an Amazon Echo Dot, can be bought for his low as $30, and you can have everything from toilet paper to technology delivered to your front door in a day. Only 10 years ago, building a new house with smart technology like fully distributed music systems in your house, lighting controls, even remote controls for your home theater system, could cost thousands or tens of thousands of dollars.

Time To Get Started

There is no better time, then right now to begin to learn how to, and implement Smart Home technology in your world. For the cost of one book (just like this one) you could start with a simple Smart lightbulb. I'm sure there's at least one light in your house that is either hard to get two or just a pain in the butt to turn. So start there.

From there you'll start to work on learning how to control your homes environment with a digital thermostat from work or the train, protect your home from fire and burglary, even start the coffee while you're in the shower. Soon every appliance in your home will be connected to let you know when the laundry is done, the food is ready, and even alert you when it needs maintenance. You probably already have almost everything you need to get started. With a smart phone or watch, wireless Internet or Bluetooth, and one digitally connected device, you can enter the world of starting to build your Smart Home one step at a time.

Why This Book?

Ronald Nutter is the perfect person to help you simplify the process and cut through the confusion. Ron helps you to understand the technologies involved, and how to easily implement them for yourself. This book is especially important because it's more like a how-to guide to integrating while utilizing multiple platforms from different companies like Amazon. Google, Apple and more.

I don't know about you, but I've often been frustrated by instruction manuals that show up with different technologies. You can think of this book as a simple guide with step-by-step instructions for each one no matter which you purchase or choose to use. From the simplest task like setting up just one smart lightbulb, to creating a smart home entertainment system, and even automating your garage or opener in appliances, this book will continue to be a reference for years to come.

I've known and watched Ron for years give simple explanations to complex situations in books, articles, and even in TV interviews for technological advice or issues like cyber security. Whether you're just considering Smart Home technology, looking to set up what you just bought, or you're ready to expand the capabilities of your home entertainment systems... This is the right book at the right time for you.

So with all that being said, I'm excited for all of us with what new technology has in store for us today, and the potential it holds for our future. To quote Star Trek, it's time "To boldly go where no one has gone before!".

Enjoy the ride!

Brian Basilico
Author • Speaker • Online Consultant

Table of Contents

Your Free Gift

As a special Thank You for purchasing this eBook, when you register your purchase at this URL http://ronaldnutter.com/homeauto, In return for doing that, I will send you a checklist of all the equipment discussed in this book. I will also email you when I have updates out to this book so that you can keep the latest copy on your Kindle device or app that you are reading it with.

As technology continues to improve and update, I will come out with update to the book, videos on YouTube and possibly a membership website to help me get the information to you in the best way possible. I will not rent, share or the information that you give me in return for signing up for my email list. Another use I can see is when I find any special deals on equipment listed in this book, I will send those out to those on my list so that you can potentially save some money if you purchase it as well. Please check out my YouTube channel for any updates to the book or new content - https://www.youtube.com/TechBytesRN.

Thank you for purchasing this book and for taking the time to read it.

About The Author

Ron Nutter has written for technical trade press since 1990. Over the years he has written for InfoWorld, LAN Times, TechRepublic, NetWare Solutions, Network Solutions and Network World. He has written and produced the HelpDesk ToolChest podcast with Ron Nutter for Network World during his time writing for Network World. In addition to that he wrote a weekly column for nearly 16 years.

He has a background in broadcasting. He assisted with the production of a live multi camera television broadcast serving in the roles of operator, director and engineer dealing with both video and audio. He working in similar roles at at a cable television company. He has written and produced several different podcasts.

Ron is also available for a limited number of speaking and consulting engagements. Register your book and get notified of updates - http://www.ronnutter.com/video101. Bookmark this link to keep up with new books as they are released - http://www.ronaldnutter.com.

QR Codes in this book

You will see a series of QR codes in this book. Whether you are reading this on a laptop or a kindle reader, these codes will give you a way to go to the information that the QR code is for. The QR code will either link to a YouTube video that goes into more detail or links to an affiliate link where the item can be purchased.

Since this is something that you may not have worked with before, I will give you the links to the resources I have used. This is a new way of making books interactive and I hope it will increase the usefulness of this book to you.

The first QR code will be for your smartphone -

Android - http://www.ronnutter.com/aqrc

IOS - http://www.ronnutter.com/iqrc

Both of these apps are from the folks at Scan, Inc. I like use the same app from the same vendor on any platform I am working on. Found that it helps minimize troubleshooting issues if they arise.

This next series of apps will be for laptops and desktops. My criteria for selecting an app for this platform was that the potential for not having a camera was a possibility. My reasoning is that if I am reading a Kindle book on my laptop/desktop that having a camera to read the QR code from the book wouldn't work since I couldn't point the camera that is built into the screen in a direction that it could read the screen. I went with apps that would essentially be a screen scraper like app that would allow me to read the code and be able to take action with the code.

Mac - http://www.ronnutter.com/mqrc ($3.99)

Windows - http://www.ronnutter.com/wqrc

Here is a test QR code to work with. It should send you to my website - http://www.ronnutter.com.

Some the QR codes or URL links in this book may be affiliate links that I will get a commission on if you buy the item. This will not increase the cost of the item you are purchasing.

How to best use this book

This book is written with two different purposes in mind. The first is to help give you an idea of the different options available to create your own Smart Home. As with traveling from one city to another, there is usually more than one way to get there, it is a matter of what works best for you.

Each section of the book gives you an idea of the different options available. The other part of each section is the troubleshooting or HowTo's that I have created as I have put my own SmartHome together. While there may seem to be some duplication between sections, there isn't. I have tried to put each section of the book together so that it stands on its own. It is doubtful that everyone will have both an Amazon Echo and a Google Home device. Instead of having a big section at the end of the book where all of the HowTo's are placed, I opted to put the HowTo's for each device in the section where I initially write about the device, so you aren't having to jump all over the book to get information on a single device.

There is no one right or wrong way to build your Smart Home. You may find that you start with one way of doing it and end up going another direction. My goal with this book is to give you an idea of what I found as I put together my version of this. I am continually making improvements to that I have in place.

Based on the information that I have collected so far, I am expecting to release another book sometime in 2019. With the way technology is changing and new technology becoming available, I couldn't create one book that had everything in it because of the way things are continually changing. Look for updates on my YouTube channel to keep up with the latest changes that I have worked with - http://www.youtube.com/TechBytesRN.

Video

https://youtu.be/ir1m69DHM2g

Suggested Order for Best Results

While it is tempting to do everything at once, I would suggest taking one thing at a time and take baby steps. Start with a couple of light bulbs and maybe an outlet. Get used to working with those before starting on projects such as going with a Smart Home thermostat or shades that open and close on command. A good way to do this is one of the starter kits that you will see discussed later in this book. It gives you the basics while letting you "sample" one way of automating your home.

The other thing to do is to take plenty of notes. Recording things such as the websites you went to for documentation, the login accounts and passwords you used for logging in to the device(s) you are configuring, etc. If you need to recreate your configuration due to equipment failure or helping a friend get their system up and running, these notes will be priceless. When you start implementing items such as smoke detectors or items that have a battery, knowing when they were put in service will help identify when they need to be replaced.

With more devices requiring some type of website login, you will start collecting more logins depending on the number and variety of devices that you will be adding to your automated home. If you aren't already using a password manager, this would be a good time to start using one. While most sites want to use your email address as your login id, you have some degree of flexibility in what you can do with a password. I would suggest that you use a different password for each site that you have to establish a login account. If one account happens to get broken into, using a different password on each site will help lessen the chance that other sites will get broken into as well.

Once you get things up and running, one site that you should look at is http:/www.ifttt.com. IFTTT stands for If This

Then That. Where this site stands out is it gives you "extensions" to the product you are tying it into and giving your functionality beyond what the manufacturer of the device released it this. You will only see a IFTTT Integration chapter for a given product if I have been able to find vendor authored services/applets/triggers/actions. This infers a certain level of support by the vendor in case you are having a problem with an IFTTT function. If the vendor authored it and/or had to create the required service to get things working, they should have some information to help troubleshoot the situation you may have to deal with at some time. While there are other applets/triggers/actions written by third parties, if dealing with something like IFTTT is new to use, I would suggest first trying the functions created by the vendor whose product you are trying to use them with as a starting point.

Smart Home's Untapped Potential

While making your own home work for you is probably the first thing on the list, there is so much more that you can do with it not for yourself but for others. Whether you have an elderly family member, an older friend who has no family, or know someone who is wheelchair bound or has other challenges, look at it from their point of view. Think of what having a Smart Home functionality can do for someone who isn't as able to be as mobile you are.

As the number of those who are getting older continues to grow, the ability to maintain a quality of life and life it on your terms will be an ongoing challenge. I am not talking about walking in and setting up everything at once. Just like I have mentioned, when helping someone who can benefit from having this kind of technology, implement it a little bit at a time. Something brand new will take some adjustment and familiarity in learning how best to work with it.

Here are some ideas to consider as a starting point. For someone who has a hard time picking up the phone or holding a cordless phone in their hand, think about what having something like a Google Home or Amazon Echo with the Connect add-on could do for them. Even if they can hold the phone, dialing a phone number can be a totally different challenge. Some the ability to dial from a pre-loaded phone book or simply speaking the phone number to dial would make someone's like so much easier. Such a small thing has the ability to make a significant impact.

Ever been some distance from home and either remembered or are not sure if you got that garage door closed ? Sometimes calling a friend or neighbor to check isn't an option. Either you have to risk leaving the door open or be uneasy about how secured your home is until you return. Later in this book, I discuss how I dealt with that exact problem. With technology that is readily available and reasonably priced, you can help someone who is older or has physical challenges be able to deal with that and have one less thing to worry about. This particular type of technology has another benefit even if someone is at home. When a service provider needs to enter the garage to get access to the basement of a house or to the HVAC system, they can now be allowed into the house without having to get up to let them in and risking the potential of a fall or other accident.

The potential doesn't stop here. With the number of smart locks for doors and sensors for windows and other points of entry to a home available, you can also help with peace of mind when someone turns in for the evening or is away from home for an extended period, that everything is locked up and safe. Your imagination will be your only limit as to what you are able to accomplish.

I ask you one thing as you read this book. Look at all of the technology I discuss here, not only for what it can do for you but what it can do for someone who can really benefit from having it available to them. Imagine how you can improve the quality of life for some who is older or has challenges to deal with that you don't have to.

If you come up with an idea of how to use something in a way that I haven't discussed or you haven't seen anywhere else, please let me know. Make sure you register your purchase of the book as a first step. As you share your ideas with me, I will in turn share it with others who have purchased the book. Whenever possible, I will also release a video on the same material. In either case, I will give you the credit. The more we can help each other, the better place this world can be.

A few years ago, this technology was as readily available or as easy to implement as it has become today. Hopefully this process will continue to become easier and offer more features as well. I look forward to being with you on this journey. Thanks for getting a copy of this book and please let me know what your ideas are for helping others.

Video

https://youtu.be/u98vNn0X8S4

Options for Home Control

While most of the items in this book have their own app for controlling the device purchased, I wanted more. That is what got me looking at voice control. For those of you who remember the Jetson's cartoon of the early 60's or the initial StarTrek TV series, you will have an idea of what I am trying to get done.

Three options will be covered in this book. HomeKit is Apple's offering. You will need an iPhone or iPad. Google Home can be very reasonable from a cost perspective if you can catch a Google Home Mini on sale for half off its normal price of $49. Amazon Alexa offering may have a little more cost to get into as to get maximum benefit, you will need an Amazon Prime subscription. Some of the cost can be minimized for startup on either the Google or Amazon solutions by using the "mini" versions of their particular voice interfaces instead of the full height versions. The tradeoff I have found with the mini versus the full height versions is the mini units were less able to discriminate between my voice and significant source of volume such as a TV. I am still experimenting with the locations I have the devices placed in so it is possible that using a different location might lessen the problem I have experienced.

Before choosing which option you want to pursue for Home Control, take an inventory of the devices that you would like to have under Home Control. Once you have that list, look at the different solutions for the devices that you want to control to see which will work best for you in terms of cost and functionality.

The HowTo or step by step instructions you will see throughout the book may be a little redundant at times but that is by design. Not knowing where you my start in the book or which parts you jump to in and in what order, I wrote the step by steps instructions so that you could start from from just about any point in the book.

Your first inclination may be to automate your whole house or apartment. I would suggest starting with one room and get things running consistently. When you start adding additional rooms, you will need to look at home your name your devices. Having a light named Chair in more than one room may be a challenge for the Smart Home to differentiate when using Voice control. This may dictate that you have a voice interface (either Alexa or Google Assistant/Home) in each room or have some ability to be to turn on a light named chair in a specific room. That is something that very well could resolve itself at some point in a seamless manner.

CHAPTER ONE

Starter Kits

Starter Kits Introduction

One way that you can help minimize the cost (at least for the initial deployment) of Home Automation is to look at purchasing a kit from a particular vendor. In theory, this should offer an assortment of the more popular items from a particular vendor. If a hub is required to talk to the different devices in the starter kit, it is reasonable to expect that it would be included in this bundle. A starter kit may be be ideal for all situations but is something worth looking at to see if any money can be saved.

As you look at the different smart home products that are available to install, see if the device in question is supported on more than one platform. If you have LED smart bulbs from Phillips that say they work with HomeKit, there is a possibility that they may also work with another platform such as Samsung's Smartthings system. While not something that you would want to do day one, it would hurt for your devices to be controlled or manageable by a different system in case you have a problem with one system that you can still use them another way.

Phillips Hue

Philips Hue White A19 Starter Kit
www.ronaldnutter.com/phsk

At the time I purchased this starter kit, I already had several different light bulbs in my setup. This one was the first device I had purchased that had HomeKit functionality. This package does come with the hub needed for the light bulbs to work. One thing to remember with any package that has HomeKit support is there will be some type of label in the box or on one ore more of the devices in the Starter Kit. As a part of the documentation you are building as a result of your Home Automation setup, this is an important item to record. Using the camera in your smartphone and something like OneNote or Evernote would be a good ay not to misplace this important information.

Samsung Smartthings

Samsung SmartThings Home Monitoring Kit
http://ronaldnutter.com/sshmk

This is the first bundle that I started with. It offers a good foundation on which to build a home automation system with. You get two door sensors, a motion sensor and an outlet module. The door sensors I later put in place on both my garage doors so that I could tell remotely which one was open or closed. While it would be ideal to be able to close the door as well, the options I found required components from different vendors and took some tweaking to get things to co-exist easily.

The motion sensor is handy for sensing when someone or something is in an area when nothing should be. The outlet module is something that should be used with care. While it is ok for turning on a tabletop radio or small TV, I would recommend caution when looking to control a larger device or something with a motor. This is where reading the specification from Samsung for this module as well as their recommendations for what should and shouldn't be connected to the module.

There are other modules from Samsung with looking at. If you have a problem periodic water in the basement or wanting to make sure that a Sump Pump is working, a moisture sensor would be good to use in this situation. One thing that has impressed me with Samsung's approach is that I am able to check the level of the battery in a particular module. These batteries will need to be replaced periodically, so being able to check the battery level remotely is a definite plus.

A side benefit that I found with the Samsung Smartthings bundle is that it gives me remote notification when the power has gone out at the house. This didn't used to be a problem for me but since the power utility that was servicing my house was bought out by a neighboring utility, power outages have become more frequent. While I may not get instant notification when the power goes out, I usually now within a few minutes. The Smartthings app on my smartphone picks up on the periodic polling from Samsung Smartthings datacenter and my hub.

One benefit that you will notice over time is that Samsung periodically pushes out updates to the Smarthub that is the foundation of their platform. So far, I haven't found a way to manually update the firmware which is probably a good thing. As a part of creating your account on their system will send you notifications when they will release updates and when you can expect to see see your hub reboot. So far, mine has rebooted during the night when it is less likely that it will be in use. As with any type of automatic reboot, there might be the occasional hiccup which might require that you manually power cycle the hub. So far, I have yet to have that happen to me.

SmartThings Link for NVIDIA SHIELD
http://ronaldnutter.com/stlns

For those that have an Nvidia Shield streaming device, Samsung has a creation option to consider. By purchasing this USB device and download a piece of software, you can now have your streaming receive also do double duty as your Smartthings hub. While not an option for everyone, it might just be the ticket for some. Depending on how much you get into Smart Home technology, it would be nice to leverage the Nvidia Shield as more than a streaming media receiver. Just to peak your interest, the Nvidia Shield also has another potential of streaming content from more than internet source such as SlingTV, Crackle and Amazon Video to mention just few of the source you will come to know and potentially use.

IFTTT Integration

To get started -

1. Download IFTTT app onto your smartphone
2. Create an email login account
3. Use a password unique to this site

Recommended ITFFF Apps to start with Samsung Smartthings

Name: Get an email when a new SmartThings Applet is

published

Publisher: SmartThings
Applet ID: dL4peb2W
Required services: SmartThings

Name: Get an email when SmartThings publishes a new trigger or action
Publisher: SmartThings
Applet ID: NyGhjenq
Required services: SmartThings

With Samsung SmartThings continually making improvements, this Applet provides a way to find out how to get better use of your IFTTT and Samsung SmartThings combination. These two IFTTT apps will help you see what becomes available as it is released. Turn things on one step at a time so that if you see any problems, you have an easier idea of where to start looking for the solution.

Troubleshooting:

1. If you are trying to turn on an applet/trigger/action and it keeps going back to off, this might indicate a problem with an underlying required. I have had this happen when trying to enable an applet/trigger/action via the IFTTT web interface. Try repeating the process using the smartphone app that you have.

2. One thing I have noticed using the smartphone app is that the first time you try to enable an applet/trigger/action that requires a specific IFTTT service that you haven't already enabled, you will be taking to a screen where you will login and authorize IFTTT to have access to base information from the service you are connecting it to. The first time you try to enable the function you are working with, it will appear to fail. What I have seen is that the service needed is getting linked but the handoff back to the function may not have

been handled the way it should have been. Try re-enabling the function and you should see it turned on at this point.

CHAPTER TWO

Lights

Lights Introduction

This is probably one of the first areas that some will start on their way to building their own Smart Home. As you will soon see, there are a variety of different smart lights to choose from. Some work directly with their own control system or hub, others can utilize your existing wireless network. There are pro's and con's to both options.

Those that leverage your existing wireless save you the potential challenge having to setup a hub to talk to the devices. The smart lights I have seen so far that use your wireless network only use the 2.4Ghz wireless band which is already getting crowded. Other systems that use a wireless hub to communicate through use a different range of frequencies that may be as congested in all parts of the country.

LIFX

LIFX Mini Day & Dusk (A19)
Wi-Fi Smart LED
http://ronaldnutter.com/lifxmdd

There are two varieties of the LIFX bulb. The first generation was a little bulky and somewhat pricey (around $50 per bulb). The 2nd generation is a little smaller and the price has dropped a bit more. You have two choices with LIFX - white or multicolor. Multicolor gives you some options in terms of doing some additional functions such as changing to blue when it is raining, flashing another color when you have a text message arrive, etc.

There will be an individual section for each method of controlling the LIFX bulbs that I have been able to setup. There will be a link at the bottom of each page linking to a YouTube video that will help as well.

LIFX Setup

Setting up your LIFX bulbs begins with creating an account on LIFX.COM and downloading the LIFX for either your iPhone or Android smartphone. Follow these steps to get that process started -

1. Create an account on LIFX
   a. https://cloud.lifx.com/sign_in
   b. Save account information in a password manager
   c. Use a separate password (and account name if possible)
2. Download LIFX app on your smart phone
3. Login to LIFX app
4. Turn on LIFX bulb
5. It will cycle several times between different colors or shades
6. When it settles down to a single color, it is ready to configure
7. Go to wireless settings on your phone
8. Look for something like LIFX Color 1000
9. Tap on that name
10. When the checkmark appears beside the network name, return to the LIFX client
11. Tap on the + in the upper right hand corner of the app
12. Tap on Connect Light
13. The light should flash a couple of times
14. Tap the Name of the Light (i.s. LIFX light 120f28)
15. Tap the gear icon in the upper right hand corner of the screen
16. Tap on Claim Light
17. Follow the prompts on getting the light to associate with the selected Wifi network

Video

https://youtu.be/HLxB1oNXTvs

Amazon Echo Integration

Follow these steps to have Alexa be able to control your LIFX lights

1. For these steps to work, you must have already have your LIFX devices working with the LIFX app
2. If you haven't already done so -
   a. Download the Alexa app onto your smartphone
   b. Sign into Alexa with your Amazon account
3. Open the Alexa app
4. Tap on the 3 horizontal bars in the upper left hand corner of the screen
5. Tap on Skills
6. Tap in the search box
7. Type LIFX in the box
8. Tap on the search icon to get Alexa to find the LIFX app
9. Tap on LIFX Optimized for Smart Home option
10. Tap on the Enable button
11. Enter your LIFX credentials as indicated
12. Tap on the Sign In button
13. When the Authorize Application screen appears, tap on the Authorize button to allow access your LIFX devices
14. Tap on the X in the upper left corner of the screen to close the page
15. Tap on the Discover Devices button to allow Alex to discover the new devices
16. Alexa should finish this process around 20 to 30 seconds
17. You should now see the new bulb(s) on the Alex Smart Home screen

In the IOS app, you can tap the icon for the bulb to turn it on, off or change the amount of dimming applied. In the Android app, you can do what you could in the IOS screen as well as control the device via voice control. This should be available in the IOS in a future update.

Video

https://youtu.be/hZs_EQYtKcg

Microsoft Cortana Integration

Follow these steps to have Cortana be able to turn lights on and off

1. If you have already done so -
   a. Download the Cortana app onto your smartphone
   b. Sign into Cortana with your Microsoft account
2. Open the Cortana app
3. Tap on the 3 bar icon in the upper left hand corner of the Cortana app
4. Tap on Connected Home
5. If Enabled connected isnt already set to on -
   a. Tap on Off to change it to On
   b. Enter your Microsoft account
   c. Tap on Ok or Login
   d. Enter the password for this login
   e. Tap on ok
   f. You should shortly be returned to the Connected Home screen
6. Scroll down the screen until you see LIFX and its icon
7. Tap on LIFX
8. Tap on Connect
9. On the next screen, you will need to enter your LIFX account credentials
10. Tap on Sign In
11. On the next screen, tap on Authorize to allow Cortana to have access to LIFX
12. You will be returned back to the screen where you tap on Connect
13. Within a few seconds it should refresh and change from Connect To Disconnect
14. Tap on the < to return back to the Connect Home screen
15. A message will show at the bottom say that it is

refreshing the devices

REMINDER: Make sure you save your account credentials for LIFX and Microsoft in a password manager on your smartphone. Try one of your LIFX bulbs you, you will say "Cortana, <name of device> light on".

Video

https://youtu.be/vNhv_gXK3hY

IFTTT Integration

To get started -
1. Download IFTTT app onto your smartphone
2. Create an email login account
3. Use a password unique to this site

Recommended ITFFF Apps For LIFX

Name: Get an email when a new LIFX Applet is published
Publisher: LIFX
Applet ID: KcegKvYi
Required services: LIFX

Name: Get an email when LIFX publishes a new trigger or action
Publisher: LIFX
Applet ID: cMXwys8i
Required services: LIFX

With LIFX continually making improvements, this Applet provides a way to find out how to get better use of your IFTTT and LIFX combination. These two IFTTT apps will help you see what becomes available as it is released. Turn things on one step at a time so that if you see any problems, you have an easier idea of where to start looking for the solution.

Troubleshooting:

1. If you are trying to turn on an applet/trigger/action and it keeps going back to off, this might indicate a problem with an underlying required. I have had this happen when trying to enable an applet/trigger/action via the IFTTT web interface. Try repeating the process using the smartphone

app that you have.

2. One thing I have noticed using the smartphone app is that the first time you try to enable an applet/trigger/action that requires a specific IFTTT service that you haven't already enabled, you will be taking to a screen where you will login and authorize IFTTT to have access to base information from the service you are connecting it to. The first time you try to enable the function you are working with, it will appear to fail. What I have seen is that the service needed is getting linked but the handoff back to the function may not have been handled the way it should have been. Try re-enabling the function and you should see it turned on at this point.

LIFX HowTo's

As with any piece of technology, there are items that are handy to know but the challenge is to keep things in a place where you can find them. The HowTo's that I have put in this chapter are ones that I collected for my LIFX devices. You may not need them frequently or at all but at least you don't have to go searching for them. My experience with LIFX has been very trouble free. There have been times that a hardware reset of a LIFX device was what was needed to get things back to normal.

Reset LIFX Color 1000 to default

1. Power bulb on
2. Power cycle bulb 5 times
3. Wait a few seconds
4. It should flash Red, Green, Blue and stop on White
5. Run through the setup procedure for the bulb

Phillips Hue

www.ronaldnutter.com/phsk

This is an affordable package to start with. You will want to select the 2 bulb and hub starter kit. The hub will need to be connected into your network using a supplied ethernet cable. While the hub does have wifi installed, it appears to be reserved solely for working with the Phillips Hue lights. Before getting to the unpacking process there are a few steps that you need to take to get up and running and quickly as possible.

Phillips Hue Setup

1. Go to https://account.meethue.com/login
2. Create a login account
3. Record the username and password in your password manager app
4. Download the Phillips Hue app to your smartphone or table

Once you have this done, get the Phillips hub connected to your network and to power. Go into Hue app on your phone or tablet and it should discover the hub shortly. If it doesn't, you may be prompted to press the Phillips button on the top of the hub. If you are trying to connect from a smartphone or tablet, you may have to try your desktop or laptop to get to the meethue.com website and get the hub registered from there. Once I did that, then I was able to see the hub from the smartphone app.

As with any piece of technology, there may be some updates that you should make sure that you have. When I first registered the Phillips hub on the website, there was an update that it wanted to push out immediately. Once I got the update in place, I was able to see the hub from the smartphone app. There was a notification that indicated that there was an additional update for the hub as well as for each of the bulbs that I had in the starter kit. It took only a few minutes for the hub to get and apply its update. The one light bulb I had running to close to 30 minutes to get an apply its update.

In the Phillips Hue smartphone client, you can check for software updates. You have the choice of doing this automatically or when you go into that part of the client. I guess from my experience with Windows, getting automatic updates is not always a good thing. To start with, I have setup my Phillips Hue bulbs for manual updates. One thing you can check in the Software Update section of the client is the release notes. This should contain information that should help you decide if you want to implement a particular update or wait until it has been around for a while. Being the first to update to a new firmware release isn't always the best option. In previous years, it was found that an update had a problem and either had to be removed or another update released to undue the problem that the earlier update had introduced.

Phillips Hue Amazon Echo Integration

1. Login to alexa.amazon.com
2. Click on Skills on the left hand side of the screen
3. Click on All Skills in the upper right hand corner of the screen
4. Type Phillips Hue in the search box
5. Click on the magnifier icon on the right side of the search box
6. Click on the Phillips Hue icon icon that should appear at the top of the search results
7. Click on Enable to activate this app for your Amazon Echo
8. Enter your Phillips Hue account credentials when the new browser window appears
9. Click Yes to grant access to Alexa to your Phillips Hue lights
10. Close the browser tab
11. Switch back to the Alexa.amazon.com browser tab if you aren't taken there
12. Click the Discover tab for Alexa to find your Phillips Hue devices

Renaming a device

1. Login to the Phillips Hue portal (https://my.meethue.com/en-us/my-devices)
2. Click on Edit Name beside the device you want to rename
3. Login to the Alexa portal (http://alexa.amazon.com)
4. Click on Smart Home
5. Click on Devices
6. Click on Discover
7. After about 20 seconds the device you updated should be available by the new name

Sonoff

Sonoff was not a device that was initially on my radar when I first started researching material for this book. The Sonoff Basic was the device that got my attention. For a $5 device, you wouldn't expect much out of it. Looks can be very deceiving. This little device can be put to a variety of uses. Even better than that is the you have the option of controlling with voice commands using your choice of Amazon Alexa, Google Assistant or Microsoft Cortana.

The Sonoff Basic is also available in a version that allows you to control it via a keychain fob device. All you need to get this up and running is a screwdriver and an extension cord that you will cut in half putting the Sonoff Basic in the middle. This gives you a "smart" extension cord for just a few minutes work.

Sonoff Wireless Smart Switch
http://ronaldnutter.com/swss

Be sure to look for another appearance from Sonoff later in this book. I will show you how to create a wireless garage door controller that will control up to 4 different doors at the same time. This is easier to setup than you might think.

Sonoff Basic Setup

1. Download eWeLink app for your SmartPhone/Table
2. Start the eWeLink app
3. Create a Sonoff account if you don't have one
4. Login with the Sonoff account if you have already created on
5. Get your Sonoff Basic and extension cord in front of you
6. Cut the extension cable somewhere between the plug and the outlet
    a. If the extension cord you are using has one wire that has some ribbing on it, make sure that if you put connect one side of the cable to the top connector on the input side of the Sonoff device, put that same "ribbed" wiring to the top side of the Ouput side of the Sonoff connector
7. Remove insulation on both wires at each end of the cable
    a. If you don't have a good wire stripper, I would suggest getting one like the Klein Tools 11054 that you will see a link to in the description for this video
    b. You shouldn't need to remove more than a 1/4 inch
    c. After removing the insulation, twist the strands of wire in a clockwise motion
    d. This will help all of the wires make it into the connector
8. Remove the covers on both sides of the Sonoff Basic
9. Position the Sonoff basic in front of you with the words Sonoff at the top of the device
10. Insert one pair of wires with the plug on it in the left side of the Sonoff Basic
    a. This is labeled INPUT on the top cover

   b. Loose the screw about 4 or 5 turns to open up the connector enough to accept the wire
   c. If you unscrew the terminal too much, it will fall out and the pressure clamp will also drop out, giving you another problem to deal with
   d. Do not loosen the screw past the point where it is higher that the terminal block it is in
   e. Make sure all of the wires from each side of the extension cable make it all the way into the connector
   f. After both wires have been inserted, tighten the screws
11. Insert the other pair of wires ending with the outlet on the right side of the Sonoff Basic
   a. This is labeled OUTPUT on the top cover
12. Tighten screws
13. Replace the covers you removed from both ends of the Sonoff Basic
   a. It is important to replace those covers as they provide a source of strain relief of the cable being accidentally pulled out of the terminals or damage to the circuit board.
14. Before replacing the strain relief covers, look closely that there arent any strands of either wire sticking out or in contact with the other wire in the pair. If there is, you have a short between the wires and sparks will fly when you plug things in.
15. Replace and tighten the screws in the covers on the Sonoff
16. Plug in the Sonoff to the wall outlet
17. Plug in a light or device to the outlet end of the extension cord
18. Press and hold the black button on the Sonoff basic for about 7 seconds (until the wifi light begins to blink fast

19. Tap the + symbol on the eWeLink app
20. If you are using android smartphone, select the first pairing option
21. For an IOS/Apple phone, select the pairing option based on how the LED blinks
22. Enter your Wifi password
23. Assign a name to the device
24. When the WiFi light is on solid, the device is now online

TIP: Do the initial setup with 10 to 15 feet of an access point. Will make the setup process goes smoother. Gives you one less thing to fight when you are using the device in what will probably be where it is installed.

Video

https://youtu.be/1tX1UBlnUA8

Sonoff After Install Checklist

1. After the Wifi light has gone solid, I would suggest cycling the device through on and off several times. You should hear a relay click each time the Sonoff Basic (or whatever Sonoff device you have just configured) between Off and On

2. Turn off the vibration

a. One thing that I found a little annoying (at least on the iPhone app) was the vibration every time the Sonoff was changed from off to on or on to off.

i. To turn off the vibration, tap on the device name (not the on or off button to the left of the device name

ii. Tap on the circle with 3 dots in it in the upper right hand corner to the right of the device name

iii. Tap on Setting

iv. Tap on the Vibration button to change it from on to off

3. Check that the latest firmware is installed

a. In my case the device came installed with version 1.5.5, the eWeLink app advised that 1.6.0 was available

b. Tap on the cloud symbol with the down arrow to start the update process

c. Tap on OK when the firmware update window appears

d. You should notice a change to the Wifi light. In my case it went to a triple flash for a few seconds.

e. After it had finished updated, the Wifi light went solid and the eWeLink app quickly updated to show v1.6.0 had been installed.

4. Document the Sonoff Device that you just installed

a. Using a label maker or a magic marker with a fine point, label the device with a unique name or the name that the eWeLink apps knows it by

b. Go into the web page for the router that you are

using and look for the page that shows the mac addresses for all devices that are connected to the router

c. You should be able to see what mac address connected to the router on what date

d. If that isn't an option for your particular router, disconnect the Sonoff device you just setup and let it stay offline for a few minutes

e. You should see a mac address either get added to the list or change state from offline to online when the device has power connected to it again

5. Surge Protector

a. Best practice for protecting the Sonoff device as well as what is plugged into it

b. More concerned about protecting what is plugged into the Sonoff

Video

https://youtu.be/wdS9LedZbA0

Sonoff Basic Amazon Echo Integration

1. Open the Alexa app on your SmartPhone/Tablet
2. Tap on the menu in the upper left hand corner of the app
3. Tap on Skills
4. Enter EweLink in the search box
5. Tap on the magnifying glass to begin the search
6. Tap on eWeLink Smart Home Fan
7. Tap on Enable to enable the skill
8. Enter your eWeLink account information
9. Tap on Login
10. Go back to the Skills page
11. Tap on the Discover Devices button

You also have the option of using Alexa to discover the devices as well

1. Alexa, discover my smart home devices
2. Alexa, discover device
3. Alexa, find device

To control the device connected to the Sonoff device you just setup, try the following commands -

1. Alexa, turn on {deviceName}
2. Alexa, turn off {deviceName}
3. Alexa, turn {deviceName} off
4. Alexa, turn {deviceName} on
5. Alexa, turn on the {deviceName}
6. Alexa, turn off the {deviceName}
7. Alexa, put {deviceName} off
8. Alexa, put {deviceName} on

Note:

1. When you add a new Sonoff based device, you will need to unlink and relink your Sonoff account in Amazon

Alexa.

2. You can give the device a different nick name when controlling it with Alexa than it is known to Sonoff. This is a good reason to document what device is know by what name and on what system.

3. All the HowTo's involving the Sonoff devices are done using the firmware shipped from the factory from Sonoff. These instructions have not been tested using any 3rd party firmware.

Video

https://youtu.be/yM9TJOh5578

Sonoff Basic Google Assistant Integration

1. If you havent already installed the Google Assistant, Google Home and Ewelink apps, do so now
2. If you have already installed both apps, make sure you have the latest versions installed
3. Go to the Ewelink app on your smartphone / tablet
4. Tap on the menu in the upper left hand corner
5. Go to Connected Home
6. Click on the Plus symbol in the lower right hand corner of the screen
7. Tap on "Smart We Link"
8. Enter your eWeLink credentials on the next screen
9. Tap on Login
10. You will see a message about linking your account
11. When that disappears, you should be able to turn the Sonoff off and on

To control the device, just say:

1. Turn on the {device name}
2. Turn off the {device name}

For examples:

1. OK Google, turn on the sink light.
2. Hey Google, turn off the sink light.

Note:

1. When you add a new Sonoff based device, you will need to unlink and relink your Sonoff account in Google Home.
2. You can give the device a different nick name when controlling it with Google Home than it is known to Sonoff. This is a good reason to document what device is know by what name and on what system.
3. All the HowTo's involving the Sonoff devices are done

using the firmware shipped from the factory from Sonoff. These instructions have not been tested using any 3rd party firmware.

Video

https://youtu.be/N8ze-fGd5HE

Sonoff Basic Cortana Integration

Since there isn't direct integration between Sonoff and Cortana, you can use a 3rd party, IFTTT (IF This Then That) to bridge between the two.

1. Create an account on IFTTT
   a. Go to www.ifttt.com
   b. Create an account on www.ifttt.com
   c. Download ifttt app for your smartphone
   d. Run the ifttt app
   e. Enter your login credentials
2. Connect Cortana to IFTTT
   a. Tap on the My Applets icon on the lower right hand corner of the screen
   b. Tap on Cortana or search for it and then tap on Cortana
   c. Tap on Connect
   d. Enter your Microsoft Cortana credentials
3. Connect Sonoff to IFTTT
   a. Tap on the My Applets icon on the lower right hand corner of the screen
   b. Tap on eWelink or search for it and then tap on eWelink
   c. Tap on Connect
   d. Enter your eWelink credentials
4. Create Applet on IFTTT to Turn on Sonoff Basic
   a. Return to main IFTTT screen on your smartphone
   b. Tap on the My Applets icon on the lower right hand corner of the screen
   c. Tap on the plus symbol on the upper right hand corner of the screen
   d. Tap + beside this on the New Applet screen
   e. Type Cortana on the search line
   f. Tap on Cortana

g. Tap on Say a Specific Phrase

h. Tap on What do you want to say

i. Type Basic on

j. Tap on Done

k. Tap on What do you want Cortana to say

l. Type Turning Basic on

m. Tap on Done

n. Tap on check mark on the upper right hand corner of the screen

o. Tap on plus sign beside that

p. Tap on Services

q. Type eWelink

r. Tap on eWelink icon

s. Tap on Turn1 channel switch on or off

t. Select the Sonoff device you want to control

u. Select on from the turn on or off option

v. Tap on Checkmark in upper right hand corner of the screen

w. Temporarily enable notification when this applet runs to help in initial debugging

x. Tap on Finish

5. Create Applet on IFTTT to Turn off Sonoff Basic

a. Return to main IFTTT screen on your smartphone

b. Tap on the My Applets icon on the lower right hand corner of the screen

c. Tap on the plus symbol on the upper right hand corner of the screen

d. Tap + beside this on the New Applet screen

e. Type cortana on the search line

f. Tap on Cortana

g. Tap on Say a Specific Phrase

h. Tap on What do you want to say

i. Type Basic on

j. Tap on Done

k. Tap on What do you want Cortana to say

l. Type Turning Basic on

m. Tap on Done

n. Tap on check mark on the upper right hand corner of the screen

o. Tap on plus sign beside that

p. Tap on Services

q. Type eWelink

r. Tap on eWelink icon

s. Tap on Turn1 channel switch on or off

t. Select the Sonoff device you want to control

u. Select off from the turn on or off option

v. Tap on Checkmark in upper right hand corner of the screen

w. Temporarily enable notification when this applet runs to help in initial debugging

x. Tap on Finish

6. Using either the Cortana app on your smartphone or the Harmon Pardon Invoke, try turning the Sonoff device on and then off

To control the device, just say:

1. Hey Cortana, Basic On
2. Hey Cortana, Basic Off

Video

https://youtu.be/CX1KhBxas7w

CHAPTER THREE

Voice Control

Voice Control Introduction

Voice control is one of the components of a Smart Home. The main thing it will do for you is to keep from having to have a basket for all of the different remote control's that you may accumulate. While Voice control isn't a requirement for a Smart Home, it will make things a little easier to work with different devices from different vendors.

There is no one right system for everyone. You may find that you may end up with a collection of devices. For some, it will be the love of experimenting with different technologies. For others, it will be due keeping the cost of implementing a Smart Home to a reasonable level.

Most of the voice control options currently available have a software client that you can download to your smartphone to try before you decide which option you want to use. There are a couple of exceptions to this, of course. Apple's Siri option is available only on the iPhone platform. Samsung's Bixby, although you can download load it for most of the Android phones, only works on the Samsung models at this point. The Amazon Alexa app is fully functional on the Android platform but wont let you access the voice control portion on the iPhone models, at this point. The video that follows will give you an idea as to what some of the potential options are in the Voice Control part of Home Automation.

Video

https://youtu.be/UHVcB_5txFs

Samsung's Bixby

Samsung's Bixby is the newest entry in the digital assistant arena. At this point, Bixby is only available on the latest versions of the Galaxy and Note smartphone platforms. Samsung has said that Bixby is not meant to be a voice control part of your Smart Home implementation.

Having said that, it does have a role. Bixby can open apps on your Galaxy/Note smartphone. For example, the vehicle I have doesn't have Android Auto or Apple HomeKit support. Using Bixby, I can open the Roav Viva app that I use for Alexa access in my vehicle. I can also open and just as importantly, close apps such as Waze for getting direction. This gets me as close as I can get at this point without having additional functionality built into my vehicle.

Amazon Echo

It seems hard to believe that the Amazon Echo started out in 2014 and was position as a smart speaker. It was thought that it would be a way to control your music and probably not much beyond that. Who would have thought that just a few years later that the Echo would become the centerpiece of a home control system. This is validated with third party products such as the Roav Viva that extends the Echo to your vehicle.

Voice Control

Amazon Echo Dot
http://ronaldnutter.com/aedot

Amazon Dash Wand With Alexa

https://www.amazon.com/Amazon-Dash-Wand-With-Alexa

Echo - Making a Phone Call

Echo Connect
http://ronaldnutter.com/ecco

1. Unpack your Amazon Connect
2. Install phone line splitter in phone outlet in wall or in the back of your VOIP adapter
3. Connect Power to Amazon Connect
4. Make sure you have the latest Amazon Alexa app on your smartphone
5. Open the Alexa app
6. Tap on the 3 horizontal lines in the upper left hand corner of the screen
7. Tap on setup a new device
8. Tap on Echo Connect
9. Read the terms on the screen and tap on Continue
10. Click on Echo Connect to place the call through your home phone line
11. The next screen advises on placing calls
12. Tap on Ok to continue

13. Tap on Continue to start WiFi setup

14. When you see orange light flashing on Connect, tap on Continue

15. Go to the setup icon on your phone and change to the Amazon-XXX

(letters may be different depending your Connect)

16. Get back into the Alexa app on your smartphone

17. The screen should now show a Connected icon

18. Select the device you want to connect the Connect to

19. Move your smartphone back to the normal wireless network

20. The alex app may show as disabled

You should now be able to make calls with Alexa Connect

The documentation says that your contact list from your smartphone can be imported. I haven't found how to do this manually. I was able to call numbers using the names in the contact list in my smartphone. The instructions say that you will need to dial numbers using a 1 and an area code. So far, I haven't had the need to put a one on the phone numbers. For some time, because I live in an area that is divided by two area codes, I have put all entries in my phone starting with the area code.

I actually had to setup my Echo Connect twice. While initial testing (outbound) showed it was working, I thought it was odd that it didn't show up in my list of connected devices in the Alexa app. Incoming calls weren't being announced as I would have expected. What finally got my attention was that the handset light was flashing for no apparent reason. I tried power cycling the unit but the problem came back. During this process, I noticed that the Alexa app had been updated by Amazon. I updated it on my smartphone. I then went through the hardware reset process by pressing the reset button and holding it for 5 seconds. I went through a re-registration process. It took about 10 minutes for a download of the latest firmware and for it to show up in my list of connected devices. A quick inbound test call now had it announcing a call as I had thought it would. All 4 lights are now on solid on the echo connect.

While it may have been unrelated to the problem just described, it is worth checking when you setup your own Echo Connect. If you have any outstanding or unheard voice messages, get those cleared from your phone or voip device. I had a couple of messages that I hadn't gone through and the signal light was flashing on all my cordless phones. While I can't definitely say that this was the cause of my problem, when I went through the re-registration process (without de-registering the device as mentioned in Alexa documentation), I had to problem going through a device reconfiguration and getting it to work as described.

With the Echo Connect properly registered, an interesting use materialized. Say your existing telephone provider (whether it is a conventional hardwired connection such as the local telephone company or a VoiP service provider such as Ooma) has a service outage of some time, with Echo connect your aren't left waiting for the problem to be fixed. Go into the Alexa app and click on Settings, Echo Connect and Outbound Calls. Change the selection to be (Alex Calling (Free). You will now be able to make calls again. Your registered mobile device with Amazon will be used as the source of caller id for outbound calls.

With the current version of firmware, there is one limitation you may run into that you should be aware of before it happens. If you call into a conference bridge that requires digits to be entered to connect you to the correct conference bridge or if you call into a company phone number that needs you to enter an extension or some other number to continue, for the time being, you will have to be creative on dealing with this. In the earlier days, you could get a small box where you could enter the dtmf tone and it played it through a speaker. If you don't happen to have one of those available, you can try first placing a call through a phone with a touch tone pad and then having Alexa pick up on the call at that point. Not ideal but you don't have a lot of options until Amazon add this ability to the Connect or Echo firmware.

Echo - Calling with Your Cell Phone

1. Make a call using the Alexa app. Before you say, why would I call with the Alexa app when I already have my cell phone handy. Good question - there are times when you are inside a building or in an area where cell phone is either questionable or non-existent. Using the Alexa app, you can call the phone number needed. When I tested this, the call arrived displaying the caller ID of the cell phone I was using. Just to make sure that it was using the wifi connection and not the cellular portion of my smart phone, I removed the SIM card.

2. If you have made the move and no longer have a landline at home (a move that I made over 10 years ago), you can extend using your cell phone as how your homebound Echo calls. I have already talked about the Echo Connect to be able to connect your Echo to your landline or VoIP provider. By adding one more item, you can have Echo call over your cell phone.

Xtreme Technologies Xlink BT Bluetooth Gateway
http://ronaldnutter.com/xtxbtg

You pair your smartphone with the Xlink BT gateway.

Connect the gateway to the Echo Connect via a regular RJ11 cable. Using Xlink BT and an Echo Connect give you the option of being able to answer the phone using a regular cordless phone with having to keep the smartphone near you.

Echo - Controlling Your Fire TV

All-New Fire TV with 4K Ultra HD
http://ronaldnutter.com/aftv

Fire TV Stick with Alexa Voice Remote
http://ronaldnutter.com/ftvs

The easiest way to have Alexa be able to control your Amazon Fire TV is the use the Alexa web interface (http://alexa.amazon.com). After logging in, you can start the setup process.

1. Click on Music, Video & Books
2. Click on Fire TV
3. Click on Link Your Alexa Device
4. If you have more than one Fire TV, select the one that you want to link
5. Click on continue if you needed to choose a specific Fire TV device
6. Select the Amazon Echo(s) that are in the same room as the Fire TV device
7. Select Amazon Apps if you would like to be able to control the TV from your phone
8. Click on Link Devices

To open a specific channel on your Fire TV, you would say "Alexa, Goto SlingTV" to open the SlingTV app on the FireTV.

To exit the channel/app you are using say "Alexa Go Home" or "Alexa Return Home"

To search for a Movie, say "Alexa search for (name of movie) on Fire TV"

To search for a particular actor, say "Alexa search for (name of actor) on Fire TV"

To watch a particular Movie/Show, say "Alexa watch (title) on Fire TV"

To pause the Movie/Show being watched, say "Alexa pause Fire TV"

To resume watching the Movie/Show, say "Alexa resume Fire TV" or "Alexa play Fire TV"

Your ability to change channels within an app will require the app supports Alexa Live voice. Here are a list of additional commands that may work for you depending the channel/app you are using on your FireTV - https://www.amazon.com/gp/help/customer/display.html?nodeId=202174250.

Note: If the title you are wanting to watch is offered free to Amazon Prime Users, it should automatically start. If it is a title you will need to pay for, you will be taken to a screen where you will make your choice of what resolution you want to view it in.

Echo - IFTTT Integration

To get started -

1. Download IFTTT app onto your smartphone
2. Create an email login account
3. Use a password unique to this site

Recommended ITFFF Apps to start with For Amazon Echo

Name: Get an email when a new Amazon Alexa Applet is published
Publisher: Amazon
Applet ID: qC6vWLcf
Required services: Amazon Alexa

Name: Get an email when Amazon Alexa published a new trigger or action
Publisher: Amazon
Applet ID: R534iJtM
Required services: Amazon Alexa

With Amazon continually making improvements, this Applet provides a way to find out how to get better use of your IFTTT and Amazon Echo combination. These two IFTTT apps will help you see what becomes available as it is released. Turn things on one step at a time so that if you see any problems, you have an easier idea of where to start looking for the solution.

Troubleshooting:

1. If you are trying to turn on an applet/trigger/action and it keeps going back to off, this might indicate a problem with an underlying required. I have had this happen when trying to enable an applet/trigger/action via the IFTTT web interface. Try repeating the process using the smartphone

app that you have.

2. One thing I have noticed using the smartphone app is that the first time you try to enable an applet/trigger/action that requires a specific IFTTT service that you haven't already enabled, you will be taking to a screen where you will login and authorize IFTTT to have access to base information from the service you are connecting it to. The first time you try to enable the function you are working with, it will appear to fail. What I have seen is that the service needed is getting linked but the handoff back to the function may not have been handled the way it should have been. Try re-enabling the function and you should see it turned on at this point.

Echo HowTo's

As with any piece of technology, there are items that are handy to know but the challenge is to keep things in a place where you can find them. The HowTo's that I have put in this chapter are ones that I collected for my Echo's. That begs the question, what do you call a family of Echo's ?

Echo General Commands

To raise Echo's volume level
1. "Alexa, raise volume"
2. "Alexa, increase volume"
3. "Alexa, louder"
4. "Alexa, set volume to 5"

To lower Echo's volume level
1. "Alexa, lower volume"
2. "Alexa, decrease volume"

Note: The volume commands can be used while music is streaming, etc.

Echo Troubleshooting

When Echo complains about not being able to connect to the wireless or not having internet access. Try the following -

1. Do a power reset on the Echo device
2. If the problem still persists, press the button with a dot on it and hold for at least 5 seconds or until you hear a message about proceeding with the use of the Alexa app
3. Proceed with updating/refreshing the Echo configuration

These troubleshooting steps also apply to the Echo Dot.

Echo Connect Troubleshooting

When Echo Connect is showing a flashing handset or is unable to make or place calls, this could indicate a problem with registration or wireless access. Try the following -

1. Do a power reset on the Echo Connect device
2. If the problem still persists, take a paper clip and insert it into the reset hold on the back of the Eco Connect.
3. Hold it for at least 5 seconds or until you see the Wireless icon is flashing orange
4. Go into the Alexa app and tap on the Echo Connect to continue the reconfiguration process.
3. Proceed with updating/refreshing the Echo Connect configuration

These troubleshooting steps also apply to the Echo Dot.

The exception to this troubleshooting process will be when multiple devices are involved. In that case, first try a reboot of your router/firewall. If your wireless access point is separate from the router/firewall that you use, that should be power cycled before rebooting the router/firewall. If that doesnt resolve the problem, do a power reset of the modem supplied by the cable/DSL/Fiber Internet provider.

Enabling Traffic Report to Work

1. Enter Alexa web portal (https://alexa.amazon.com)
2. Click on Settings
3. Click on Traffic
4. Click on Add Address
5. Enter Work address
6. Click on Save Changes

To ask for a traffic update, try one of the following

phrases -

1. How is the Traffic?
2. What's my commute?
3. What's the traffic like right now?

Enabling Multi Room Music

1. Enter Alexa web portal (https://alexa.amazon.com)
2. Click on Settings
3. Click on Multi Room Music
4. Click on down arrow beside group name
5. Select a group name (in my case, everywhere)
6. Check the boxes besides the Echo's that you want in this group
7. Click on Create Group button

Command to use multi room music

1. Play "Artist Name" everywhere
2. Play "Song Name" everywhere

To stop the music streaming, say "Cancel"

Notes:

1. If the Echo where you start the music streaming from loses power, the music doesn't always start playing on that specific echo on reboot
2. When you want to stop the music streaming, I have found that you have to initiate that process from the Echo that was used to start the music streaming.

Personalizing Echo

By default, each Echo device is assigned a name that generally includes the first name of the person who bought it. One option for changing the name is to put the name of the room of the room as the name of the device. It will help keep the locations of the devices a little easier to keep track of.

1. Go into Amazon Alexa app
2. Tap on 3 bar menu icon
3. Tap on Settings
4. Tap on Echo device that you want to rename
5. Tap on Edit beside Device Name
6. Enter new name of Device
7. Tap on Save button
8. New name of device should be active shortly

Another change to consider making is to change the wake word used when you have multiple Echo's present in close proximity to each other. For example, I have an Echo in my living room and an Echo Dot in an adjoining room. When I am in the right area, I can bring up both Echo's at the same time.

1. Go into Amazon Alexa app
2. Tape on 3 bar menu icon
3. Tap on Settings
4. Tap on Echo device you want to change the wake word on
5. Tape on Edit beside Device Name
6. Scroll Down to Device Name
7. Tap on down arrow beside current wake word
8. Scroll through the list of currently supported wake words
9. Tap on the desired wake word that you want to use
10. Tap on Done
11. Tap on Save

Amazon Wand

Amazon Dash Wand With Alexa
https://www.amazon.com/Amazon-Dash-Wand-With-Alexa

The Amazon Wand is Amazon's way of making it easier for you to shop and purchase items from Amazon. It appears to have an unintended benefit as you can use it as a portable Echo when you need to do in the basement and want to do something such as turn on lights or other items that you have configured in your Smart phone.

This little device has a lot of potential. While it cannot play music, it is wide open to being used as a source of voice control. For less than $20 US, you have an interesting way to start getting experience with using voice control with your Smart Home. For those times when you would like to have an Amazon Echo or Echo Dot nearby and maybe don't have a power outlet handy, the Amazon Wand might just be an option.

Amazon Wand Setup

To setup Amazon Wand -

1. Open a web browser on your smartphone/tablet
2. Go to http://www.amazon.com/wandsetup
3. Click on US Dash Wand Setup
4. Click on the Get Started button
5. Install the batteries as directed and click on Continue
6. Turn on Bluetooth on your smartphone/tablet
7. Press and hold the button on the wand unit the light turns orange
8. Click on Connect on the smartphone/tablet
9. Once the wand is setup
10. Select the wifi network your smartphone/tablet needs to connect to by tapping on the name. If you have both 2.4Ghz and 5Ghz SSID's, you will only see the ones that run on 2.4Ghz.
11. Enter the password for the SSID in the password box
12. Leave the save password to Amazon option checked
13. Click on the Continue button
14. Once the device is registered, the light will to from orange to green
15. You will hear a chime come from the wand
16. If you try to scan a barcode soon after it is registered, you may get a message that the device is updating. Try again in about 5 minutes or so.
17. The light will probably turn white at some point during the update process
18. After that it will turn green and then eventually to out.
19. The Wand should be ready to use at this point
20. Turn off bluetooth on the smartphone/tablet
21. Now is the time to run a few tests
    A. Press the button and scan a barcode
    B. You should get a message that the item has been

added to your cart.

C. Press the button and ask Alexa what the time is

D. You should get a response with the current time

22. Your wand is ready to start using

A few items to remember -

1. Anytime you want the Alexa wand to do something for you, you will need to start by pressing the button

2.Try all of the commands you use with your regular Alexa. The only commands I know of at this time that don't work involve playing music, listening to the radio or making a phone call with Echo Connect module.

3. If you try pressing the button without the wand being registered, you will be told to go to www.amazon.com/-setup to setup the device. Don't bother, the link that the Wand is giving you doesn't work.

4. Don't be concerned with the Wand not showing up in the Alexa app

The Amazon Wand comes with two AA alkaline batteries. I normally use rechargeable batteries whenever possible. See how often you have to replace the batteries in your wand as to whether you need to consider using replaceable batteries or not.

Amazon Wand Troubleshooting

Wand says "I am having problems understanding right now"

This can mean one of two things. The bar code you are attempting to scan is bad or invalid. You can try scanning the bar code about one to two inches between the bar code and the wand.

Wand says "Added"

This an indication of a good scan. The item you just scanned should be in your Amazon card shortly.

Wand says "I am a little tired right now. Could you try again later."

This is an indication that it is unable to scan the barcode you are trying to have it read.

Wand light on or flashing

The Wand has information for you concerning the battery level, setup information or another message that you need to hear. Press the button to hear the message.

Google Home

Google Home appeared on the scene in 2016 just two years after Amazon's entry into the market. You will see two software components to the Google Home experience. The piece that you will interact with the most when setting up your Google Home environment is the Google Home app. This is where you will initially set up the Home, Home Mini and Home Max hardware components. Linking your smart home devices is another area where the Google Home app comes into the mix.

The Google Assistant app is your gateway using voice commands to getting information from the web instead of using a keyboard. It also serves as the voice interface to turning lights and other devices off or on. Depending on the features of a particular device, you can perform additional steps such as dimming a bulb or changing the color it uses.

Google Home Setup

1. Download the Google Home app for your smartphone
2. Unpack and plug in your Google Home Mini to AC power
    A. A series of 4 dots will start flashing
    B. When Google Home is ready, you will be told to download the Google Home app to your smartphone
3. Turn on wifi on your smartphone
4. Open the Google Home App
5. If you have already setup a Google Home device, you should see a tile that says a new device has been found
6. Click on setup in the tile announcing the new Google device
7. When the Google Home App has connected to the new Google Home app, you should hear a sound from the Google Home device
8. Tap on Yes to continue
9. Tap on the Room that the Google Home will reside in
10. Tap on Next
11. Accept the supplied name or change it to what you would like it to be. This can be changed at a later time.
12. Tap on Next to continue
13. Tap on the Wireless network that you want it to connect to
14. Tap on Next to continue

15. You will see a series of 4 dots rotating until it has connected

16. Tap on Next to continue

17. Tap on Next to continue

18. Tap on Next to continue

19. Tap on next to continue

20. Tap on Continue to proceed

21. Verify address presented an tap on Next to continue

22. Tap on Next to continue on Add music services screen (these can be added later)

23. If Cast devices are found, they will be listed and you will be given the option of adding them to your experience. Tap on Next to continue

24. When the Almost done screen is presented, click on Next to continue

25. When Google Home is ready screen appears, Tap on Continue to proceed

26. When the "Just start" screen appears, tap on Finish Setup

If you are using something other than a gmail account and your account is hosted using Google Docs service, you will need to have the administrator for Google Docs service make a few setting changes to allow things to work with Google Home - https://www.theandroidsoul.com/enable-google-assistant-apps-work-allo/

There are some additional changes/adjustments you will need to make such as training Google home to recognize difference voices to personalize the experience.

Google Home Mini

1. Download the Google Home app for your smartphone
2. Unpack and plug in your Google Home Mini to AC power
   A. A series of 4 dots will start flashing
   B. When Google Mini is ready, you will be told to download the Google Home app
   to your smartphone
3. Turn on wifi on your smartphone
4. Open the Google Home app on your smartphone
5. Click the Get Started button on the lower right hand corner of the screen
6. Select the email address used to associate the Google Mini with
7. Click on the OK button on the lower right hand corner of the screen
8. If you get the message about error connecting to the cloud, click Try again and you
   should be able to continue
9. If you don't have Bluetooth turned on, enable it
10. If Bluetooth is already enabled, your smartphone will start searching for the mini
11. When it is found (it should be something like GoogleHome8409), click next to
    continue
12. You should hear a sound once the two devices are

connected

13. Click on Yes to continue
14. Select a location for the Mini
15. Click Next to continue
16. Choose the Wifi network to connect the Mini to
17. Click Next to continue
18. Enter the Wifi password
19. Click on Connect
20. You will see a Connected message when the Mini has connected to Wifi
21. Click Next to continue to setup the Google Assistant
22. Follow the onscreen prompts and you should be up and running shortly.

If you are using something other than a gmail account and your account is hosted using Google Docs service, you will need to have the administrator for Google Docs service make a few setting changes to allow things to work with Google Home - https://www.theandroidsoul.com/enable-google-assistant-apps-work-allo/

There are some additional changes/adjustments you will need to make such as training Google home to recognize difference voices to personalize the experience.

Google Home - Making a phone call

You can call using your Google Home device by using one of the following commands -

Call by business name - "Call The Pizza Place"

Ask for nearest business, then call - "Where is the nearest restaurant ?", "Call Them"

Call by number - "800-555-1212"

Call by name in your contact list - "Call Parents"

To hangup, do one of the following -

"Hey Google, stop"

"Hey Google, disconnect"

"Hey Google, hang up"

"Hey Google, end call"

Remember that when making a call with Google Home, you aren't able to make a call to 911 for the time being. Making calls to international numbers and 900 numbers are only possible if your Google Home has been linked with Project Fi (Google's cell phone service) or Google Voice (if you have dialing credits available).

One thing that I tried was to use the Google Assistant app to make a call. To make sure it would go out over Wifi, I removed the SIM from the phone. When I tried to make a call, I was greeting with the error "Not Registered on Network". It looks like, at least for the current time, that calling over Wifi using the Google Assistant app isn't supported. Don't worry, there is another Google app to the rescue.

Google Voice is an option to have for those situations when you may not be able to make a call. You will end up having another phone number but if you are in a situation where you existing cellular provider and/or phone doesn't support calling over Wifi and you need to make a call, this is an option to consider. Google Voice is an app or service that is in a state of transition. Google Voice is being made a part of Google Hangouts. When I tried to make a call using Hangouts, I wasn't able to find the right way to do it. Found out that I needed a Google Hangouts Dialer app. The native Google Voice app is still here and still works but in case it does go away at some point, I wanted to make sure you were aware of the next step.

Google Home HowTo's

As with any piece of technology, there are items that are handy to know but the challenge is to keep things in a place where you can find them. The HowTo's that I have put in this chapter are ones that I collected for my Google Home devices. You may not need them frequently or at all but at least you don't have to go searching for them.

Google Home Troubleshooting

When Google Home complains about not being able to connect to the wireless or not having internet access. Try the following -

1. Do a power reset on the Google Home device
2. Go to the Google Home app on your smartphone
3. Look for a tile that shows your Google Home device as being ready for setup
4. Tap on the square to continue and follow the steps given to you

The exception to this troubleshooting process will be when multiple devices are involved. In that case, first try a reboot of your router/firewall. If your wireless access point is separate from the router/firewall that you use, that should be power cycled before rebooting the router/firewall. If that doesn't resolve the problem, do a power reset of the modem supplied by the cable/DSL/Fiber Internet provider.

Google Home Voice Troubleshooting

When Google home starts complaining about not being able to recognize your voice, try these steps to resolve the problem -

1. Do a power reset of the Google Home device
2. Go into the Google Home app
   A. Tap the menu icon
   B. Tap on Devices
   C. Tap on the 3 horizontal dot icon beside the icon for the Google Home device
   D. Tap on Settings
   E. Tap on More
   F. Tap on Voice Match
   G. Tape on Teach your Assistant your voice again option
   H. Follow the instructions on the smartphone screen

Google Home - IFTTT

To get started -

1. Download IFTTT app onto your smartphone
2. Create an email login account
3. Use a password unique to this site

Recommended ITFFF Apps For Google Home

Name: Get an email when Google Assistant published a new trigger or action
Publisher: Google
Applet ID: cMXwys8i
Required services: Google Assistant

Name: Get an email when Google Assistant Applet is published
Publisher: Google
Applet ID: t3QPRrE
Required services: Google Assistant

With Google continually making improvements, this Applet provides a way to find out how to get better use of your IFTTT and Google Home combination. These two IFTTT apps will help you see what becomes available as it is released. Turn things on one step at a time so that if you see any problems, you have an easier idea of where to start looking for the solution.

Troubleshooting:

1. If you are trying to turn on an applet/trigger/action and it keeps going back to off, this might indicate a problem with an underlying required. I have had this happen when trying to enable an applet/trigger/action via the IFTTT web interface. Try repeating the process using the smartphone

app that you have.

2. One thing I have noticed using the smartphone app is that the first time you try to enable an applet/trigger/action that requires a specific IFTTT service that you haven't already enabled, you will be taking to a screen where you will login and authorize IFTTT to have access to base information from the service you are connecting it to. The first time you try to enable the function you are working with, it will appear to fail. What I have seen is that the service needed is getting linked but the handoff back to the function may not have been handled the way it should have been. Try re-enabling the function and you should see it turned on at this point.

What is Android Wear?

You may have heard of this and you may not. Basically, Android Wear is a version of Google's Android operating system that is designed for use on Smartwatches and other wearables. So far, I have only seen Smartwatches use this but other things will come in time.

When you start looking for a Smart Watch, there is one thing that you will want to watch is what version of Android Wear that the device will support. You may find some Smart Watches at almost a giveaway price. Check the specifications very closely for something called Android Wear 2.0. This is the latest version of AOS (Android Operating System) that will run on some but not all SmartWatches. The manufacturer of the watch, in most cases, will need to be the one that releases the latest version of Android Wear.

Android Wear has the potential of being an extension of allowing access to your Google Home implementation to your watch. Depending on the particular app you are using, you may experience limited abilities to control your smart home devices depending on the app. For example, one app that I use has limited abilities to control lights. Instead of being able to control individual lights, you only have the option to control lights that are parts of a scene (a collection of devices). With the Google Assistant feature that is a part of Android Wear 2.0, it is just like have a Google home device but on your wrist for having access to voice control options.

Selecting a Smart Watch

As with any item there are multiple sources to choose from. A while back, I started with a Motorola Moto 360 watch running Android. When I tried to revive it for this book, I ran into a challenge. Motorola had decided that it wouldn't offer the latest version of Android Wear. It may have been due to the hardware not supporting it. I wasn't able to tell for sure. So I started looking for a Smart Watch that would be able to run the latest version of Android.

Asus ZenWatch 2
Smartwatch 1.63"
http://ronaldnutter.com/azw2

I found that ASUS was one potential source. There were two versions of the ZenWatch to choose from. Trying to keep things on a budget, I went for a refurbished version of the ZenWatch 2. After doing a lot of research, I found that the ZenWatch 2 would support Android Wear v2. Once it arrived, I let it charge for several hours before doing anything. Not being sure of how long the battery would last, I didn't want to cut it too close.

To get starting with the whatever Smartwatch you choose, you will want to install the latest versions of the Android Wear - SmartWatch and the Google Assistant apps. If you already have the Google Assistant app installed, then just make sure it is on the latest version. If you go with the Asus

ZenWatch 2 or 3, you will want to install the latest version of the ASUS ZenWatch Manager app so you can do some additional watch face changes as well as suggestions for ASUS ZenWatch specific apps.

Microsoft Cortana

Microsoft entry to the Smart Home arena started in 2014 at the Microsoft Build 2014 developer's conference. Initially it was released for Windows 10 and Windows phone platforms. For those that love the history of things, knowing the background of things can be just as interesting. Initially, Cortana was just the code name for the service that was demonstrated. Due to the significant response from those that saw the service, the codename became the name.

Later, Cortana would be available on both iPhone and Android devices. It also offers the ability to track flights, set reminders and timers, getting the latest news and so on. Not to be outdone by the Echo and Google home platforms, you also have the ability to have Cortana in a smart speaker form. Harmon Kardon and Microsoft partnered to make the Invoke that and more.

Where Cortana is playing catchup is in the area of Home Automation/Control. While it doesn't have as many individual options as competing products, using a product such as Samsung SmartThings to act as a bridge to those devices. As with any piece of SmartHome technology, it is growing and adapting to the changing environment.

Setting up Cortana on your Smartphone

1. Create a Microsoft.com account
   a. If you don't have an account, click on Sign In
   b. Enter your email account
   c. Click on Next
   d. Click on Create Account
   e. Use a password for this service/site that you aren't using anywhere else
   f. Enter your login credentials into a password manager
2. Download the Microsoft Cortana app from the app store for your smartphone
3. Enter your Microsoft login credentials when you start the Cortana app

Note: Depending on how well you type on a smart phone, it may be easier to create the account using a desktop/laptop browser

Video

https://youtu.be/S3gxX9JO6no

Setting up the Harmon Kardon Invoke

1. Download the Microsoft Cortana app to your smartphone
2. Turn on Bluetooth on your smartphone
3. Open the Cortana app
4. Tap on the 3 horizontal icon to access the menu
5. Tap on the Devices option
6. Tap on the dot beside "I accept Harman/Kardon" license agreement
7. Plug in the power for the Invoke
8. Tap on Next
9. When the lights on the app match what you see on the top of the Invoke, tap the Next button
10. Tap on Yes to link the Harman Kardon Invoke to your Microsoft account
11. Tap Yes to grant the necessary privileges to your Microsoft account
12. Tap I Agree to acknowledge the last screen of account permissions
13. Tap Ok to grant Cortana to access the contacts on your smartphone
14. Confirm your location and timezone on the next screen
15. Tap on the Confirm button at the bottom of the screen
16. Tap on Next to continue
17. Tap on the HK Invoke name on the next screen to continue the linking process
18. Once the app has "found" your wireless network, it will display that network and allow you to change the name of the device
19. Tap on Next to continue
20. The screen will indicate when the Invoke has joined your wireless network
21. You hear a message that the Invoke needs to get an update

22. Tap Done to continue

23. Tap Next to continue when you see the Almost there screen

24. When the "Pick a preferred music provider" page appears, select the provider and Tap Next

25. Tap Done

26. When the "Connect your calendar" page appears, tap Skip to continue if you don't need to connect to your Office 365 account if you have one

27. Review the video and then tap on "I've seen it" at the bottom right hand corner of the screen

28. When your Invoke has finished getting any updates it needs, you will hear an audio message indicating that

29. You can tap Skip to bypass the messages that give you suggestions on what you can do with your Invoke. If you are new to Cortana or the Invoke, go through the messages, you might learn a few ways to use it that you hadn't thought about.

Go to this URL to get the latest on Cortana's skils - : microsoft.com/en-us/windows/cortana/cortana-skills/

If your Cortana didn't come with any manuals, go to the HK website to get the manuals -

https://www.harmankardon.com/on/demandware.static/-/Sites-masterCatalog_Harman/default/dw71dcb1aa/pdfs/Harman%20Kardon%20Invoke%20Owners%20Manual.pdf

Grab the quick install guide and spec sheet for the Invoke while you are there and store them in OneNote so you will always have them handy.

Video

https://youtu.be/q-gUXO5JzF0

Siri and HomeKit

For voice control of your HomeKit devices, at this point in time, you have one choice - Siri. That is good and bad. It would be nice to be able to tie in an option such as Alexa or Google home but that isn't in the cards at this point.

In another chapter, I shows how to setup the Apple TV (at least 3rd gen from what I have read - I have a 4th gen Apple TV). In the latest iPad, there is an option of using it as a HomeKit hub. That is an expensive way to have a HomeKit hub when you compare it to the cost of a 4th Gen Apple TV at $150 US.

Depending on what type of accent that you have and how you pronounce certain words, you may have to adjust how you name your devices. Even in a quiet room, I have experienced issues with Siri on the iPhone 6 Plus that I have being able to reliably distinguish between chair and share. Hopefully future improvement to HomeKit and/or Siri will help increase the reliability of being able to distinguish between words.

Something else to remember with HomeKit that is useful. Just because the box the Smarthome tech that you have purchased may only say HomeKit, don't let that make you think that is your only option. I bought some Smarthome devices that said they supported HomeKit. Nowhere on the box did it mention that it supported the other technologies such as Z-Wave, Zigbee, etc. It was almost by accident that I looked up the name of the vendor in the list of supported devices of one of the Smarthome hubs that I already had in place. Even better was that I found that the device I had would respond to more than one platform. From experience with other technologies, I thought that once you had selected a particular platform that you were locked to the platform until you did a factory reset. While this may not be the case with all of the smart tech platforms but it does give you some hope for some degree of flexibility. One thing that you might notice is that if you turn on a LED smart bulb using HomeKit and turn it off, say using the Samsung Smartthings platform, that the HomeKit interface might not realize for a while that the device in question has been turned off using another option. This may not happen but it is a possibility.

Enable HomeKit on iPhone

1. Tap on Settings
2. Scroll down to the Home Icon
3. Tap on Home
4. Tap on cloud icon to download the Apple Home app

The Apple Home app should now be on your iPhone

These are the steps you can expect to go through when adding a device to HomeKit

1. Tap on Apple Home app
2. Click on Get Started button
3. Click on Allow on the next screen so you can control items when you aren't at home
4. Click on Add Accessory button
5. Using camera, scan the HomeKit code in the box the for device you want to control
6. When the next screen appears, select the item shown there
7. It will be added to HomeKit momentarily
8. Click on the Next button to continue
9. Click on the Next button to continue
10. You should now be on the Default Home screen

HomeKit setup tips

One of the things that makes HomeKit easy to setup for the first time Smart Tech user is the special "key" that is in the box that your HomeKit compatible Smartthings devices are packaged in. This key is something that you will want to have a backup copy of. Whether you use something like Evernote or Microsoft OneNote, take a picture of this code with your smartphone or tablet. I have not implemented that many different types of HomeKit at the point I am writing this portion of the book but check that code with each additional item of HomeKit compatible hardware that you purchase. Whether this code stays the same or differs, having a copy of the setup code in a place where you can find it will certainly be better than having to keep all of the boxes of that technology you buy, in case you need it.

Event if you don't need the HomeKit setup code today, especially if aren't implementing HomeKit today, having it may be beneficial down the road. Before you get too heavily invested in one platform, being able to "kick the tires" or test another platform might be nice before you choose one way of doing things when another way might have been a better choice. It might be interesting to have a device reachable by more than one smartphone platform, so that you can really amuse your friend by using one method to use a device to start and then another to stop or turn off the device.

CHAPTER FOUR

Home Entertainment

Home Entertainment

While not one of the first areas that a Smart Home incorporates, it will probably be one of the more heavily used. The streaming media receiver is becoming more and more popular since the momentum of moving from a satellite receiver or digital cable box to other options is increasing. My journey began when my fiber internet provider left me down for a week because they were only able to come out from 8 to 5. After I got the internet connection restored, I put in a backup connection from another provider as a backup plan.

If you are like me, I grew tired of continually being in the middle between a provider and one of the sources that I got content from because one side was supposedly being unreasonable in negotiations. With having a streaming media receiver, I now have the option of changing providers with a few keystrokes. When writing this book, I experienced an unintended benefit. Hallmark additional a 3rd channel to their current channel lineup. I didn't have to do anything to get the 3rd channel, it just appeared one day.

You have your choice of 4 different streaming receivers that I will discuss in this book. There are more than that available. They are usually based on the Android TV OS but may be on some type of custom firmware. There is no one right choice that can be made. One of the first things I would suggest is looking at the content available that you might be interested in and let that be one of the things that guides your decision. One of the first areas that I used when selecting a media receiver was a particular type of technical training that I was looking for. Not all of the training providers had training "channels" available on all of the platforms.

Home Entertainment is not limited to just streaming media receivers. You also have the option of adding "smarts" to your existing television without paying a lot of money for something that may only be supported for a few years. The Logitech Harmony Hub is an interesting device that can, for example, turn on the device and switch to a desired input or channel. With a little bit of work, you can even add voice control to that using your choice of Amazon Alexa or Google Assistant. I haven't had a chance to try integrating this with Microsoft Cortana but I suspect that it should be possible to get it done.

Amazon Fire TV

Amazon Fire TV
http://ronaldnutter.com/aftv

I have had an Amazon Fire TV box for several years. Until recently, I had used it mostly for seeing the "free" movies or television shows that were offered as a part of the Amazon Prime subscription program that Amazon is well known for. Having said that, to fully get the use out of this box, you will need to be an Amazon Prime member. You can check this link for the current listing of channels available on the Amazon Fire TV. You may also see the service referred to as Amazon Video on some hardware platforms such as the Roku for example. Here is a link to see what channels/ services are available on the Fire TV Platform - Amazon Channels - www.amazon.com/channels.

Amazon Fire TV Stick
http://ronaldnutter.com/ftvs

Fire TV - Updating Firmware

1. Click on Settings
2. Select and Click on Device
3. Click on About
4. Select and Click on Check for System Update

If display changes from Check for System Update, to Checking Now and then back to Check for System Update, there is no update to be applied.

Apple TV

Talking about the Apple TV wasn't one of the things I had planned for this book. I had tried an Apple TV several years ago and hadn't been that impressed. In building a list of the channels/services that I wanted to watch, that is what showed me that I needed to have the Apple TV in this book. One of the Technical Training channels that I will talk about later in this book had the Apple TV as one of the few streaming platforms that it supported.

Getting the Apple TV up and running doesn't take that long. It takes longer to remove all of the protecting wrapping that is on the device and get everything out of the box. You will want to make sure you have a HDMI cable that is long enough to run from the TV to where you want to put the Apple TV device. While I would have preferred to have purchased the 4K version of Apple TV, since I don't have a 4K TV at this point and there was additional cost involved (only Apple charges extra for 4K), I went with the cheaper option.

One thing that is a little different with Apple TV is the remote. You can't change the battery in the remote like you can with the other streaming devices. Remember that USB to Lightning cable that came in the box with Apple TV ? You will need that to charge the remote. Plug in one end of the cable into the remote and the other into the charger that you use for your iPhone. Allow about 2.5 hours for a full charge. If you are impatient like I am, you will find that the remote has some charge out of the box so you can start using it as soon as you unpack your Apple TV.

One of the first things that I would recommend doing as soon as you have your Apple TV connected is to do a firmware update. While, by default, this will happen automatically, I prefer to start with the latest firmware from the start. Depending on when you receive your device and the speed of your internet connection, the initial update may only take a few minutes.

The Apple TV remote takes a little getting used to. The "gesture" area acts something like a trackpad. If you have a long email address or password, you will become very familiar with how to use it. The "gesture" area also does doubles duty as what I will refer to as a mouse clicker or how you select what you want to do with the Apple TV. I have found one nice feature with the Apple remote that varies a bit depending on the streaming channel or service that you are using. When in SlingTV, swiping up from the bottom shows you the current channels in your subscription and what is currently showing. Swiping down give you the title and details of the show you are currently watching and the ability to turn on subtitles.

When the Apple remote gets low on power, you should get some type of prompt on the TV to let you know that the remote is at 20% or less remaining power. You can go ahead and plug it in to charge. When the remote sees that it is plugged in, you should see a message indicating that. The missing link at this point is that you don't get a corresponding message when the remote is fully charged. This may seem like a minor thing but I don't like to keep a device plugged into a charger to avoid damage to the battery or otherwise shortening it's useful life.

Apple TV - https://hd-report.com/apple-tv-channels/

Apple TV Remote Shortcuts

Here are a few "shortcuts" for lack of a better word that I have found in using my Apple TV Remote. There is one things that we need to remember when dealing with the remote. That is the different between a tap and a click. A click will be when you press hard enough in the top part of the remote that you hear it click. A tap will be a similar gesture but you shouldn't feel the click of the remote. There is a third gesture that you will use is a swipe. A swipe will be similar to a tap but your finger will move across the top area of the Apple remote but no click should be felt.

Single Tap - will display the time line of the video being shown. The number all the way over to the right will be the total length of the current video.

Single Click - when watching a video, this will pause what you are watching. Single clicking again will resume the current video.

Double Tap - if the video being watched has a black band at the top and bottom of the video, doing a double tap on the remote will bring the video full screen. A second double tap will return the video back to its original format. If this function doesn't work, the current video source doesn't support it.

Down Swipe - this will cause a drop down area to appear on the screen. What you will see initially is more information about the video being watched. Depending on what other options the video supports, you may have the option of turning on subtitles or other features.

Up Swipe - this is a way that you can get the information that appeared on the screen to go off the screen. Depending on the video source you are using, such as a service like SlingTV, you may have the option of seeing a thumbnail version of the current program guide. Down Swipe to clear this off of the screen.

Apple TV HowTo's

This section covers a variety of tasks for your Apple TV. This is to help you keep things up to date, add new channels, etc. Whether you have one device or several, the how-to's in this section will help make the process easier to keep things running to best level possible.

A part of of this section of the book is a general troubleshooting guide. As you experience builds with the equipment that you will be acquiring, you will start making your own additions to this guide with items unique to your implementation. Who knows, you might end up being the streaming media guru for your friends!

Updating Firmware

1. Click on Settings
2. Click on System
3. Click on Software Updates
4. Click on Update Software
5. If an update is available, click on Download & Install

Adding a channel

1. Go into App Store
2. Search for Sling
3. Select and click on Sling TV
4. Select and click on Get
5. Click on Get
6. Open Sling TV Channel
7. Enter credentials for service

Some channels/services may give you the option of entering your credentials for that service or take a code that the Apple TV will generate and you enter that into a specific web URL for the service you are wanting to link the Apple TV to. This certainly beats having to enter the credentials using the Apple Remote and the swipe gestures to step through the letters, number, etc that are in your login name and password.

Checking Battery in Remote

1. Click on Settings
2. Click on Remotes and Devices
3. Click on line with Remote and Battery logo
4. Look for line showing Battery Level

How to reboot/restart

1. Click on Settings
2. Click on System
3. Click on Restart

Controlling Jumps in Audio Level

There are times that some of the content I listen/watch can jump without warning. Not all content on the Apple TV has this option but it is worth looking for if you need it.

1. Swipe down on the Apple remote
2. Select Audio
3. Select Reduce Loud Sounds
4. Swipe up to return to the content being watched

Using your iPhone/iPad as a Remote

1. Download Apple TV Remote app from the App Store
2. Make your the Apple TV is up and running
3. Tap on the Apple TV Remote app
4. Tap on Add Apple TV
5. Tap on the Apple TV that is displayed
6. Your TV will display a 4 digit code
7. Enter the 4 digit code in the Apple TV Remote app to pair the devices

Harmony Hub

Logitech Harmony Hub & Remote
http://ronaldnutter.com/lhhar

The Harmony Hub is a device sold by Logitech. When it is setup, it gives you the ability to have one "system" that will control multiple devices. What it also does is give you an option of pressing one button and have it turn on a TV to a specific channel. There is a version of this that has just the hub (the flat black box on the left side of the picture above). I bought the full package that included the remote.

Depending on the type of smartphone you have, there is an app that you can download to be used in place of the remote or as a part of the configuration process. You will need to create an account, either using the app or via the Logitech website. There will be 3 options for creating an account - Facebook, Google or Email. I would suggest using an email account. The main reason is that don't like to give access to one site using credentials I use for another. As with any new account I create, I would use a password unique to the Logitech site that isn't used for any of the other logins that you have. If you haven't already started using a password manager, now would be a good time to do so. As you create more accounts, having a good way of keeping track of the usernames, passwords and links to get to a specific site will save some hassle later on when you need access to the site for software upgrades or to make configuration changes. In case you misplace the documentation that comes with the Hub, here is the link to the site you will need to use - https://www.myharmony.com/en-us/.

Another benefit to the Logitech Harmony system is that it can integrate with either Google Assistant/Home or Amazon Alexa. The first thing to do is get the Logitech Harmony running on its own before connecting it to something else. Keeping track of the configuration you have created would be a good thing to do as a part of this process. While it may be a hassle while you are getting the device set up, it can save some problems later on trying to figure out how something was setup when you are trying to figure out what it isn't working.

Harmony Hub Setup

1. Download the Harmony App to your smartphone (there is a PC and Mac version that can be used as well.

2. Launch the app and select Set Up New Hub

3. Enable Bluetooth on your smartphone or tablet

4. Ensure your smartphone or table is connected to your wifi network

5. Make sure the Harmony hub is powered on

6. The app should find your hub shortly

7. Once the hub has paired with the phone over Bluetooth, you will be asked to enter your Wifi credentials

At this point, you will need to login to the app using the account that I suggested be created earlier. Follow the instructions on the screen and you will get walked through the remainder of the setup. I would suggest just setting up your TV at this point. You can add the other devices later. Take things one step at a time.

Harmony Hub - Amazon Echo Integration

1. Login to alexa.amazon.com
2. Click on Skills
3. Type Harmony in the search box
4. Press Enter
5. Click on the skill named "Harmony" authored by Logitech
6. Click on the Enable button
7. Click on Email
8. Enter email address and password for your Logitech Harmony account
9. Click on Sign In
10. Click on Authorize to enable Echo to access your Logitech Harmony settings
11. Click on the right arrow on the screen until you get to a screen that has "Link Account" at the bottom
12. Change back to the tab containing alexa.amazon.com
13. Do a discovery of devices and you should see the Harmony Hub appear

Commands:

To Mute TV - "Alexa, ask Harmony to mute TV"

To Unmute TV - "Alexa, ask Harmony to unmute TV"

To pause Movie being watched, "Alexa, ask Harmony to pause on TV" or "Alexa, Pause"

To unpause Movie being watched, "Alexa, ask Harmony to play on TV" or "Alexa, Play"

https://support.myharmony.com/en-us/harmony-experience-with-amazon-alexa

Harmony IFTTT Integration

To get started -

1. Download IFTTT app onto your smartphone
2. Create an email login account
3. Use a password unique to this site

Recommended ITFFF Apps to start with for Harmony Hub

Name: Get an email when a new Harmony Applet is published
Publisher: Harmony
Applet ID: arBxVZdF
Required services: Harmony

Name: Get an email when Harmony publishes a new trigger or action
Publisher: Harmony
Applet ID: HSNX8c5H
Required services: Harmony

With Logitech continually making improvements, this Applet provides a way to find out how to get better use of your IFTTT and Logitech Harmony Hub combination. These two IFTTT apps will help you see what becomes available as it is released. Turn things on one step at a time so that if you see any problems, you have an easier idea of where to start looking for the solution.

Troubleshooting:

1. If you are trying to turn on an applet/trigger/action and it keeps going back to off, this might indicate a problem with an underlying required. I have had this happen when trying to enable an applet/trigger/action via the IFTTT web interface. Try repeating the process using the smartphone

app that you have.

2. One thing I have noticed using the smartphone app is that the first time you try to enable an applet/trigger/action that requires a specific IFTTT service that you haven't already enabled, you will be taking to a screen where you will login and authorize IFTTT to have access to base information from the service you are connecting it to. The first time you try to enable the function you are working with, it will appear to fail. What I have seen is that the service needed is getting linked but the handoff back to the function may not have been handled the way it should have been. Try re-enabling the function and you should see it turned on at this point.

Nvidia Shield TV

Nvidia Shield
http://ronaldnutter.com/nvsh

The Nvidia Shield device wasn't something that was initially on my radar for inclusion in the book. That was mostly due to it's price (nearly $200 US) and there are only so many streaming hardware devices that one can look at when you are buying each one that you are working with. Due to a bargain I found on Black Friday and some interesting information I found about the Shield being able to do double duty as a DVR (Digital Video Recorder) got me to take a second look at it.

The Shield runs on what is known as Android TV. This put the Nvidia shield in a little different category than the other streaming devices that we have discussed so far. By adding a wireless keyboard/mouse and installing a web browser on the Shield, you have a pc capable of surfing the net. So at this point, you have a device that can receive streaming media, function as a DVR and can browse the internet. If that wasn't enough, there is one more function you can task your Nvidia Shield to do which is controlling your light, smaller appliances and a few other things. I will cover that subject in another book.

There are two different versions of the Nvidia Shield. One come with the remote control you see just to the right of the Shield picture above. The other version comes with the gaming controller that you see at the far right of the same picture. If you plan on using the Shield for just streaming media, then the remote control only version will be the one to go with.

Starting up the Nvidia Shield for the first time was a very interesting experience. It leverages your Google account, so if you don't already have one, get one. Since I have the enhance level of account protection enable on my Google account, I expected some challenges with getting things setup. The Shield didn't miss a trick. Guess that speaks to how much thought appears to have been put into the Android TV OS. Unlike some of the other streaming devices I have worked with, the Nvidia Shield automatically downloaded the latest available upgrade as a part of the initial install process.

As I started working with the Nvidia Shield, I found another benefit that I hadn't noticed before. Several years ago, I got one of the early Google Chromecast. It was somewhat of a hit and miss situation. It worked but seemed a little kludgy in its operation. The biggest drawback was that if I forgot I was "casting" from my laptop to the Chromecast and shut down the browser, what I was watching also stopped. One of the features of Android TV OS is that it also comes with Chromecast functionality. I didn't have to do anything to setup it up. The first time I tried to cast something, I remembered that I had unplugged the chromecast. Imagine my surprise when I saw "Shield" in the list of devices. Even better was when I exited the browser tab I had open and the content I was watching continued uninterrupted. Looks like somewhere between the Nvidia

Shield and Android TV OS, Chromecast really came into its own. Glad I happen to see that before I ordered the latest generation of the Chromecast dongle.

Nvidia Shield HowTo's

This section covers a variety of tasks for your Nvidia Shield. This is to help you keep things up to date, add new channels, etc. Whether you have one device or several, the how-to's in this section will help make the process easier to keep things running to best level possible.

A part of of this section of the book is a general troubleshooting guide. As you experience builds with the equipment that you will be acquiring, you will start making your own additions to this guide with items unique to your implementation. Who knows, you might end up being the streaming media guru for your friends!

Updating Firmware - Nvidia Shield

1. Click on Settings
2. Select and Click on About
3. Select and Click on System Upgrade
4. Select and Click on Check for System Upgrade

If display changes from Check for System Upgrade, to Checking Now and then back to Check for System Update, there is no update to be applied.

Adding a Channel

1. To go the main Nvidia Shield screen
2. Select and click on Google Play Store
3. Highlight and click on microphone in upper left hand corner
4. Speak name of what you are looking for (i.e. YouTube)
5. Search results will be display on screen
6. Select and click on the app/channel you want to install

Roku

Roku 2 SX

I have had a Roku 2 SX device for several years and when I started writing this book, I knocked the dust off of it and was curious if it was still working. To my amazement, it was and there was a significant firmware upgrade available to apply. Once it had completed the upgrade process, I looked for a Google Play Movies app. After installing that app and configuring my credentials, I was happy to see that a movie I had purchased via the Google Play Store was there and waiting to be played.

Roku has done a good job at continuing to support devices that are several years old. Not everyone trades up to the latest device every year or so. It is refreshing to find a company that has a good customer service attitude. I had not used my Roku 2 SX device for quite some time and was very happy to find that a major update was available for it.

The number of channels is being added to the Roku platform grows a little bit each day or so. When look at what version of the Roku platform to purchase, look for a checklist of features supported by each platform to see which version may be the best for you. At the time I began writing this book, I was looking at versions of the full appliance, namely the Premier Plus and Ultra models. I usually take a look at different reviews on a given problem to see what others have experience. While at the time I was look at the difference between the two models, I did see something that caught my eye. While today, the Ultra might have more processing power than what is needed, there could be another advantage that hasn't made itself apparent yet.

The FCC recently approved, by a narrow margin, an update to how content is broadcast from a conventional TV station. While that applies to just TV stations at this point, it could impact streaming services in the near future. The newer your streaming hardware is, the more likely it will be kept up to date or problems fix than a device made four years ago or longer might. Time will only tell on this but having a newer version of a device at least gives you a better chance of firmware updates being made available than an older device might give you.

Roku Premiere Plus

http://ronaldnutter.com/rpp

Roku Premiere Plus

After getting my feet wet looking at the different streaming services available on Roku, I was curious if the newer hardware offered any improvements worth looking at. I had thought about going with the latest available hardware, which in this case would be the Roku Ultra. At the time I looked for it, it was a challenge to find a place that had it in stock. It was good luck or good timing that Black Friday was just around the corner, I found the Premiere Plus for just under $50 with free shipping.

I anxiously awaited the delivery of my Premiere Plus curious to see what improvements might be in store. There were several improvements worth noting. Moving through the different channels on the Roku took far less time. There is now a screensaver option available with, of course, a selection of screensaver to download (both free and paid). The thing I was most curious about about how the Premiere Plus would perform over a series of hours of continuous use. What I am referring to is something we have all done at some point otherwise known as binge watching. I picked a series that I had only see an episode or two when it had first aired. I proceeded to watch 6 seasons for several hours over a period of days. The box got the slightest bit warm but no where near what I was expecting.

Another nice feature I ran found when I registered my latest Roku was that during its initial setup all of the channels I had installed on my older Roku showed up on the Premiere Plus with no action required on my part. The Premiere Plus is definitely running on a faster processor as I didn't notice a slowdown in menu response like I did with the latest code when I upgraded my Roku 2SX. I have noticed on the Premiere Plus device that when I am watching a TV series that I am usually give the option of automatically viewing the next show in the series (unless I am at the end of the season) or stopping where I am. On the Roku 2 SW, I have to back up a level in the menus and manually select the next how segment to watch. Admittedly a minor feature but one that I have enjoyed immensely in the short time that I have been able to use it.

If you are setting up a Roku for a family member or have a Roku that you want to have a different "personality" or set of channels from the other Rokus, set this device up with a different username and password. If you only have one email account, see if you can setup an alias email address so you can have what appears to be an additional email account without having to physically check an additional email account. Make sure that you mark or label this Roku in a way that you know it was setup to use a different email account to avoid some confusion at a later point or when you need to get some support for this specific device.

Please check this link for the latest channels available for the Roku - https://channelstore.roku.com/browse.

Roku HowTo's

This section covers a variety of tasks for your Roku. This is to help you keep things up to date, add new channels, etc. Whether you have one device or several, the how-to's in this section will help make the process easier to keep things running to best level possible.

A part of of this section of the book is a general troubleshooting guide. As you experience builds with the equipment that you will be acquiring, you will start making your own additions to this guide with items unique to your implementation. Who knows, you might end up being the streaming media guru for your friends!

Roku Setup

1. Unpack the Roku you have purchased
2. Connect the Roku to your television using a high-quality HDMI cable
3. If possible, connect your Roku to your internet connection using a network cable
4. Plug in the power supply to the Roku and into an available AC outlet
5. Install batteries into the Roku remote
6. Turn on the TV and select the appropriate HDMI input
7. Select the appropriate language to using with your Roku
8. Wait for the Roku to update to the latest available firmware
9. Create a Roku account and activate your Roku device
    a. Use a unique password for this login
    b. Enter this into a password manager

Updating Firmware - Roku

1. Click on Settings
2. Click on System

3. Click on System Update
4. Click on Check Now

Apply the update if one is indicated for your system. The Roku will apply it and then reboot.

Plex Media Server

This is one area that I hadn't really thought much about. To say that Plex is a Media Server doesn't really do it justice. Plex gives you the ability to view your photos, movies and television shows in one console. What you are looking at here is what the business world would refer to as client/server technology. The Plex server is the part that is the repository of the media that makes up the pictures, movies and television shows that you have recorded. The client is the app that you run on a streaming media server. That app is what talks to the one or more media servers that house your media collection.

As you get more involved in the Plex system, you will start hearing the term "transcoding". Basically this is how the Plex server "converts" the media file(s) it has to be displayed on a different type of type of display (smartphone, tablet, etc). Where this becomes an issue the number of different devices that will be playing streams at the same time. The more this happens will impact the resources of the Plex server. The load of transcoding the media, recording and playing OTA (Over The Air) broadcasts from local broadcasters can be overwhelming to some hardware. This is something that you need work on one step at a time to get a good idea of what the Plex server is capable of. As tempting as it is, enable one feature at a time will be the best way to learn how to deal with issues if they come up because you will have a better understanding of how things are working.

Video

https://youtu.be/MEyVRCyPe2A

Plex Server Platforms

You have three different options for the platform to use for your Plex server. There is a choice of Windows, Mac, Linux and Nvidia Shield. Each platform has its pros and cons. One way of deciding which platform will work for you is to look at what experience you have with each of the platforms I have mentioned. Windows and Mac are probably the two most popular options.

I started of with using a Mac Mini. It have several options working for it. In my case, I had a Mac Mini equipped with 16 Gigabytes of RAM and a 1 TB hard drive. The advantage in this was that I had a self contained system to start with. I decided to run my Mac Mini in a headless configuration. That means that it had no locally attached monitor, keyboard or mouse. I enabled remote access on the Mac Mini which allowed me to use VNC on another computer to access my Plex Server. Since there is the occasional power outage, I also have the Mac Mini set to automatically power up when it senses power. The option to have Plex automatically start up when the Mac Mini boots up is also selected so that Plex will be available as soon as the server starts up or restarts after applying a patch.

Using a Windows or Mac platform as a Plex Media server allows you to do other tasks with the server while it running the Plex Server application. One platform that has been a little more challenging is the Nvidia Shield. Unless the Plex Server app is running in the foreground, the Plex.TV portal and the Flex app on a media streaming server wont see the server as up and running. My experience with the Plex app for the Nvidia Shield is that it is mainly a server app and has limited client functionality. It has potential, so it is one that I will continue to work with.

Regardless of the platform, you have a license issue to consider. While it is possible to run a license free version of

Plex, there are benefits to going with the Plex Premium license. The feature that I have use the most is Live TV which allows you to view your favorite shows from the Plex server. I will cover that in more detail in another chapter. Plex DVR (Digital Video Recorder) allows you to record the TV shows and movies that may be playing at a time when you won't be at home and watch them at a more convenient time. The Premium license also includes the ability to use Photo Albums to organize your collection of pictures into easier to view categories.

Video

https://youtu.be/W4TCoVYOJfo

Plex Live TV / OTA

This is one of the first features I tried from what the Plex Server has to offer. It has been one of the more frequently used options. To enable this feature, you will need to have the PlexPass Premium license active. While you are trying this and some of the other options that require the PlexPass to work, you can start with a 3 month license for a reasonable price. Once you have made the decision to work with Plex on a more permanent basis, you can go to either a yearly or perpetual license. One of the benefits of a PlexPass license is that you have a electronic program guide (EPG) that can be downloaded on a daily or more frequent basis. It is tailored to the method you are using to receive local broadcasts.

The first component required to enable Plex Live TV to work is a tuner. There are a variety of tuners that Plex supports. I chose the HDHomeRun Connect tuner to work with. One benefit is that was a self contained device that could work independently of the different platforms (Window, Mac, Linux) and didn't require any special drivers to be installed. There are several different models you can choose from. One decision you will want to make is to decide on how many different programs you want to record at the same time. The Connect Duo model contains two tuners. This will allow you to record two programs at the same time or be recording one program while you watch another live. The more people that in the same household will impact on the number of tuners you may need to have before having to make a decision on how may programs can be recorded at the same time.

The good news is that if you start out with the Connect Duo and need to have a little more flexibility on recording an additional program, you can always add an additional tuner to your Plex Media server. While there is no published information that I am aware of that says you can use different brands of OTA (Over The Air) tuners available, it is suggested that you use the same brand (and model if possible) of tuner to minimize any potential tuner related issues.

If you already have a tuner that isn't on the supported list, you may find a way in the Plex forums to enable it to be used. In that case, it will be best effort support and you may find little to no information from Plex Support if you run into problems. In the short time I have worked with Plex, I have found it to be a versatile and powerful platform.

When implementing an OTA tuner, you will need some type of antenna. Using sites such as www.antennaweb.org and http://www.antennasdirect.com to find how many stations are in your area, the direction and distance they are from you. The two choices in antennas that you have to work with are inside and outside. Inside antennas may be easier to work with but may not pick up as many stations as their outside counterparts. You may find that moving the antenna as little as a foot or two may make a significant amount of difference in the quality of the signal you receive.

If you end up going with a secondary tuner, you will want to look at the trade off of using a secondary antenna or using an antenna splitter. The issue may run into with a signal splitter is that the single level may drop just enough to cause you to not receive one ore more stations that you could receive with just one tuner and antenna. One thing that you can use in place of an antenna splitter is a signal amplifier. At a minimum, this should give you the same signal level at both amplifier outputs. This is another area where going with cheap cable and/or connectors could cause additional problems which might end being even more challenging to resolve.

To find out which combination of antennas, cabling and amplifiers will get you the best signal level will require you either purchase a piece of test equipment made of this purpose which you will probably only need during setup or to use what the SiliconDust HDHomeRun tuners incorporate into their line of OTA tuners. After having conversation with the tech support folks at SiliconDust, I got some good information that I wanted to pass onto you in this book.

The three signal measurements are Signal Strength, Signal Quality and Symbol Quality. While these may sound like the same time, there are some differences to be aware of. Think of Signal Strength to be how loud the signal is at your location. The last two measurements Signal and Symbol quality are the two to watch the most. Signal Quality is how clear the signal is or when thinking of in terms of listen to a radio, how clear you hear what is being said. The Symbol Quality refers to how much of the information you could hear clearly or guess. Where the guessing comes in refers to the error correction that is in the signal and if it was able to correct for errors in what was received.

The easiest way to get to the information is to download the HDHomeRun signal meter apps from with the Apple store or Google Play store. When looking at several different channels, you may find that you need to exit and restart the application. By the time you read this it may no be an issue for you or may be related to the combination of the app and the smartphone hardware that I used when I tried the apps. In either case, this gets you to some good diagnostic information that you might not know about any other way. The more tools you have at your disposal will make the job of getting the best signal levels possible a little easier.

Video - Setup of Connect Duo

https://youtu.be/RBTXjFWV0xY

Video - Setup of Plex Live TV

https://youtu.be/meeTWyhnoTM

Plex - How much storage to I need ?

This, as they say, is the 64 million dollar question. The best answer I can give you is that it depends. Plex Live TV DVR will probably use the most storage. The average number I have found on my Plex implementation is 2 GB of storage for a half hour show and 4 GB of storage for an hour long show. These number may change for you depending on what you are recording, the signal quality, etc. The number for pictures/photographs can vary a bit more depending on the resolution of the camera used and what is in the picture. Using these numbers as a rough guideline, you should be able to take the different types of media that is in your collection to have an idea as to how much storage you will have to end up with once you have everything moving into digital form.

As you get started with your Plex server, a 1 TB drive may take you a while to get full. As you have needs for storage beyond a single TB (TerraByte) drive, you will want to look at something called a NAS or SAN. The main difference is that a NAS is usually single device or chassis containing multiple drives. A SAN is local network containing multiple storage chassis contains one or more drives each.

Before you decide to go buy a NAS device, take a look at the NAS devices that are currently on the support list on the Plex website. As you centralize all of the different types of media you have, start thinking about what you will do if a drive containing your media were to fail. Even with the quality of the drives currently available, it is reasonable to expect they will run for quite some type but as some point they will fail. The 3 types of RAID (Redundant Array of Inexpensive Disks) that you will encounter in this type of situation are Raid 0, Raid 1 and Raid 5. There are other levels of raid but typically require more sophisticated hardware than you will find in a home type NAS. Raid 0 is where data is striped or spread across multiple drives. The disadvantage here is that, in most cases, when a single drive fails, it is likely that you will loose the entire set of drives or at least the data on the drive that failed. Raid 1 is limited to two drivers. One drive is a mirror image of the other. In this case, one of the two drives can fail and you still have your data. That need to happen at that point is to replace the failed drive. A disadvantage is that you can't increate the amount of storage in a matched drive set unless the manufacturer of the NAS has built in the functionality for you to upgrade the drive size. Raid 5 gives you a little more protection from drive failures. The minimum number of drives in this Raid level is 3. For example if you have 3 1TB drives, you will have 2 TB of usable storage. One drive can fail and you are still up and running. Some NAS systems offer the option of a hot spare which would be a fourth drive that would be idle unless a drive failed and then the information on the failed drive would be reconstructed from the data on the remaining drives that are still operational.

In an ideal situation, when you find out that you need to add more storage, you should be able to add an additional drive. Another option would be to upgrade 1 drive in a RAID 5 set, let the rebuild process occur, change out the next drive and repeat this process until all the drives have been replaced with higher capacity ones. There might be some additional work to do to use the additional storage but will depend on the particular NAS system that you are using.

The thing that is hardest on a NAS is the startup process or an upgrade shutdown or loss of power. Depending on how often you have a total loss of power or periodic brownouts, you may want to think about putting at least the NAS on a UPS (Uninterruptible Power Supply) to keep the dips or surges in voltage from doing damage to the NAS device. If you haven't worked with a UPS before, it is basically a battery where the DC voltage is converted to AC that needed by the device(s) that are plugged into it. The size of UPS you need (how long it will run before the battery is completely discharged) will depend on the power requirements of the devices that are plugged into it and how long you want it to be able to run until power is usually restored. One thing to think about is that over time, the battery in the UPS wont last as long during a power outage, so at some point it will need to be replaced. At a minimum, you should consider putting the NAS on a UPS as well as the Plex Media server.

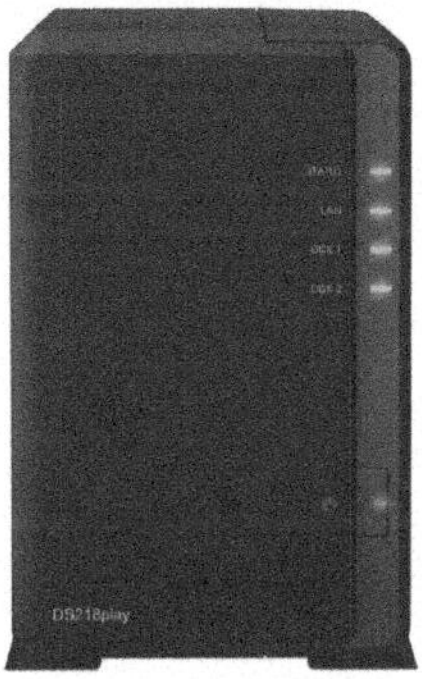

Synology 2 bay NAS Disk Station
http://ronaldnutter.com/s2bnas

WD Red 2TB NAS Hard Disk Drive
http://ronaldnutter.com/wdr2tb

Setting up a Plex Scanning Station

To convert your existing collection of slides, negatives and photos, you will need to setup some type of scanning station. If you only have slides to convert from analog to digital, you can look for a simpler type of scanner. If you have both type of media to convert, the best scanner for you may take a little more research.

Just as with a car or truck, there are quite a few choices to work with. This is another area where going cheap could cost you a lot of time in having to rescan item you had already scanned because of a poor quality image. While you are looking at scanners, see what type of software that comes with the scanner. Ideally the scanner you choose should have some software with it to help you get started until you find scanning software that is better suited for the type of scanning you will be doing.

Epson Perfection V600
http://ronaldnutter.com/epv600

The scanner that I went with was the Epson Perfection V600. While can spend more on a scanner, it may be money better spent elsewhere or not at all. One of the things that got my attention was that the V600 came with film and slide carriers so that you could scan slides and film strips (negatives and positives). Since I had quite a few carousels of slides, being able to scan 4 slides at a time would definitely help speed up the process. Being able to get a good scan of the slide was important because some of the slides I have are old enough that the color is well on the way to fading out. There is a light source in the lid of the scanner that you use when scanning slides or film.

The software that comes with the scanner has been great to use so far. Some of these slides that haven't faded have shown signs of some type of color shifting. One of the options in the Epson software allows for me to do a certain amount of color correction. For it being just a checkbox option in the scanning process, I have been pleased with how well the images look after the scan. For the pictures that were taken in a backlit situation, there is also an option that just might make a particular slide or picture go from being useless to one that can bring back memories for years to come.

If you can, I would suggest using the computer that you will be using as your Plex server to be your scanning station computer while you are scanning the photos and slides into digital form. When you setup your Plex server, it may have created a series of directories - one for movies, one for TV Shows (if you are using the Live TV option) and another for pictures/slides. You can think about using the existing directory structure and then creating subdirectories for each type of media that will be scanning in. What I am doing is that I create a directory for each tray of slides that I can in. For the collection of albums containing pictures, I will follow the same process. As I scan in each type of media, I create a separate subdirectory and library entry in Plex. The main benefit from doing this is that you wont have to scan the media on one computer, copy it to a flash drive and then copy the flash drive onto the Plex server.

Plex Voice Control Options

As with other parts of this book where I talk about voice control for light bulbs, garage doors, etc., controlling your Plex server is no different. The only limitation is that you will have to use a Roku streaming media device. While researching this part of the book, I found three different options. One is from the folks at Plex themselves and the other two options are from a 3rd party.

Plex - Amazon Alexa Skill

This voice control option was created by the Plex folks themselves. It was easy to implement. For this skill to work, you will need to enable Remote Access on your Plex portal for the Plex server you are working with. This will also involve doing port forwarding in your firewall to allow a specific port to pass through. If you aren't used to working with firewalls, you might need to find a friend or someone who is familiar with how to do this. The documentation or online help for the firewall you are using should help walk you thought the process. The main thing I would suggest is to use a port number other than the default one listed. To use a port number such as 22381 or similar high port number to one that isn't commonly used by other services or apps.

Once you have that working, the next step will be to go to your Alexa portal (http://alexa.amazon.com) search for that skill. During the process of enabling this skill, you will need to enter your account credentials on Plex with Alexa. Before enabling this skill, if you are using the same password on Amazon as you are on the Plex portal, I would suggest changing it before linking the accounts together. I haven't had a linked account compromised yet but I like to take every reasonable precaution to keep there from being a problem.

For this skill, this is the only thing that you need to do to get it to work. In order for it to work, I have found that you need to be in the Plex app before things will start to work. There are no subscriptions charges for this to work.

Video
https://youtu.be/wifgcp57N3w

Quick Remote for Alexa & Roku

This app takes a little more to get it to work and has a little more functionality. You will need to download the app from either the Apple store or Google play store depending on the smartphone that you use. The app needs to be linked to the Roku that you are using and to Alexa. This solution from Appestry gives you voice control over your Roku to the point that you can tell it to open the Plex app and keep going from there.

This is the first app that I have found where there is a charge for using it past a certain point. You are allowed to use up to 50 commands per month before you need to have a subscription for things to continue to work. There are several videos that show you how to get things working.

I found a different way that worked as well. The first thing to do is have your smartphone on the same wifi network as the Roku. Open the Quick Remote app and connect to your Roku. The next step is to link the Quick Remote app to Alexa. You will need to grant the Quick Remote app the privileges list on the Alexa screen.

Quick Remote for Google Assistant and Roku

For this app to work, you will want to have the Google

Assistant and Google Home apps installed on your smartphone. As with the Alexa version of the app, you will want to make sure that wireless is enabled on your smartphone and it is one the same SSID as your Roku. Tap on Select Roku to get your Roku connected. Tap on the Sign into Google option to link to your Google Account. You will want to try one of the commands possible. This will help complete the last part of the process where you will get a card in the Google Home app to complete the configuration process.

One challenge you may run into is where is speaking going on from the television and get either Google Assistant or Alexa to hear you. If your Google Assistant or Alexa device aren't near you and are closer to the television, this is more likely to happen. If you have the Google or Alexa app on your Android smartphone, you may have a way around that issue if you encounter it.

Harmon Kardon Invoke

Harman Kardon Invoke
Voice-Activated Speaker with Cortana
http://ronaldnutter.com/hkinv

Initially, I was sent the Harmon Kardon Invoke as a part of my work with Cortana Voice Automation. Using a speaker that I connect to via Bluetooth isn't something that I had really thought about. That changed the other day when I had some music that I wanted to listen to when I was driving to and from work. Listening to it using my Samsung Galaxy S9+ just wasn't going to cut it.

It must have been fate. I looked to where the Invoke was sitting and saw the Bluetooth button on the base of the Invoke. Pairing it couldn't have been any easier. Since I have an ASUS Zenwatch connected to my Galaxy S9+, Bluetooth was already turn on. All I had to do was press the Bluetooth button on the Invoke. Within seconds, the invoke told me what device to select on my Smartphone. The two devices paired almost immediately.

If I had been in my vehicle, I would have immediately started playing music. I realized that I didn't have anything loaded on the phone. Since I am an Amazon Prime member, I had a whole host of music at my beck and call. I downloaded the Amazon Music app. As soon as I started the app, it confirmed the username I wanted to login with and all my music was there. As soon as I picked some music to play, it immediately start playing through the Invoke without my taking any further steps. Even having to select the Invoke the first time I used it would have been understandable but I didn't have to do that. That is the way technology should work!

Even with just the basic connection that I had just built, I was happy with the quality of sound I heard from the Invoke. With the Harmon Kardon name, I would not expected any less. I hadn't seen everything yet, with the Galaxy S9+, I had another card to play. Go to the Setting menu to start. Scroll down to the Advanced part of the screen and look for Sound quality and effects. Tap on that option to continue. You will next be at a screen labeled Dolby ATMOS. Tap on the slider to turn it on.

There are a few more things to walk through and you will be very happy with what you will hear. Scroll down the screen until you see Sound Personalization. Tap on Adapt Sound. On the Adapt Sound screen, you will have a selection to make. By selecting your age, an optimization will be put in place that will help you get the best sound from Invoke tailored to you specifically.

Start playing some music and I think you will be happy with the result. Try turning off the ATMOS function to hear the different that it makes. If you play a particular type of music most or all of the time, there is another change you can make on the previous screen under the Equalizer settings to set it to the type of music you listen to.

Video

https://youtu.be/q-gUXO5JzF0

CHAPTER FIVE

Other Smart Home Areas

Garage Door

This was one the areas where I needed some help the worst. It was also the area that took the longest to find a solution that was the most cost effective and easy to implement. As with anything relating to technology, the option that is best today could be different tomorrow. I moved from a residence that had no garage doors to one that had two. Combine that with having a family member who occasionally needed to come to my house to pick up my dog when I was working late made the ability to monitor the garage doors to make sure they were closed helpful.

While this option may seem the easiest, it will also take a little more work. If your garage door opener was manufactured after 2003, you just might be able to find solution fairly easier, maybe. The first thing to do is to get the EXACT model number and manufacturer of your current garage door opener. You will need that to see if the "smart" garage door controller will be able to work for you

Monitoring Door Status

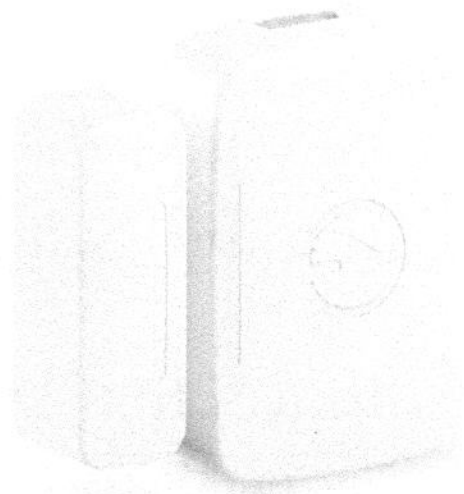

Samsung SmartThings Multipurpose Sensor
http://ronaldnutter.com/ssmps

I first found this solution as a result of the Samsung SmartThings Starter Kit that I purchased. The Multipurpose Sensor has a feature that makes it ideal for being used as a way of monitoring if a Garage Door is open because of it's ability to sense if it is titled or otherwise not in a vertical position. Due to the adhesive pads attached to both the sensor and the magnetic part of the sensor make it easy to put the sensor in place or move it without having to drill holes. This option makes it an ideal option for use by renters.

The SmartThings smartphone app makes it easy to quickly check on the status of both the garage doors. I have both door sensors tagged as favorites so that all I have to do is start the SmartThings app and know within seconds the status of both doors. With one tap on a particular door, I can quickly check on the condition of the battery. That certainly beats having to remove the battery from the sensor and check with a volt meter.

Although not a requirement, I usually do a test of both doors while raising and lowering them. While the doors are in motion, the status is checked to make sure it goes from closed to open while raising the doors and from open to closed when lower the doors. One reason to check the activity of the sensors is when there is a significant drop in temperature so make sure the switch hasn't hung in an open or closed state.

This was the first solution that I found that was reasonable to implement in terms of cost and easy to implement. If you find that your current garage door opener has the necessary features to be integrated into your smart home, having a backup system to be able to tell if the doors are really closed might be good to have in place.

Controlling Your Garage Door

This is where you will really need to do your research. Having the information on your existing garage door opener will be required to have a chance to getting the best solution in place. If you haven't already decided on what voice control option you want to use, make sure that supporting all of the current voice systems (Alexa, Google Assistant and Cortana) can be used with whatever smart home garage door controller you end up choosing. Even with that option you may need to be careful. Investigate closely to what level each of the voice control systems is supported. You may find that Google Assistant may be supported out of the box but other voice control options may require an ongoing subscription to a service to be supported. Looking at a third party option such as IFTTT that has been used earlier in this book may be the intermediate option to have in place until full support is available at a future date.

If you have more than one garage door that you would like to control, that is another item that should be on your checklist of things to look at for each option. Some options require that a component be purchased for each garage door. Other options can support two or more doors. Even if you have more than one door, having multiple control options might be a better choice since if a single controller were to fail, you would still have control over the other door(s) that are on a separate controller.

Something else to watch out for is that Wifi is supported as a connectivity choice. This helps open the possibility that control by voice is an option. If Bluetooth is the only option, this means that using your smartphone will probably be the only way you can remotely access your garage doors.

Sonoff 4 CH Pro Smart Switch v2
http://ronaldnutter.com/s4chprov2

This is an interesting option that I almost missed. In doing my research for the best smart controller solution, it was starting to look like it might be easier to install a new garage door opener. When I was looking for an option that could possibly work with my current garage door opener, I started looking for a relay that would make and break a connect in about a second. My thought in this case was to "fool" the garage door controller in to thinking that the button had been pressed on the controller when it was actually another device that was performing that action.

The plus with the Sonoff solution is that it contains four relays on the inside of the housing that can control four different doors in the case I am looking to use it. The other feature it has is that it supports something called "inching". A better way of describing this is that it can create a connection, hold it for a specified period of time and then release the connection. That is exactly what happens with the control panel for the garage door opener I currently have does.

NEXX Garage NXG-100

http://ronaldnutter.com/nxg100

The NEXX Garage option is one of those that has the ability to be supported by multiple voice systems. The NXG-100 is one of the options that requires a unit to be purchased for each garage door that will be controlled. The plus here is that it is supported by multiple voice options without requiring an ongoing subscription in order to use some of the voice control options that you may be considering.

As I found out, having the specifics are very important. With the NEXX option, I found out that my particular door opener wasn't on the supported list. This is where closely reviewing the company website provided information indicated that in those situations where a door opener wasn't supported they were offering an add-on item that might just get you a working solutions (and them a new client).

Setup of Sonoff 4CH Pro Rev2

Follow these steps to get your Sonoff configured to control your Garage Door

1. Download eWeLink app for your SmartPhone/Table
2. Start the eWeLink app
3. Create a Sonoff account if you don't have one
4. Login with the Sonoff account if you have already created on
5. Install a power cord on the Sonoff 4CH Pro Rev2
6. Turn Bluetooth and Wifi on your Smart Phone
7. Plug in the Power cord for the Sonoff 4CH Pro Rev2
8. Once the blue led starts blinking, hold down one of the buttons for about 7 seconds
9. Tap on the plus sign in the eWelink app
10. Tap on Next
11. Enter the password for the Wifi SSID you are connecting to
12. Tap on the Next button
13. Tap on OK to grant eWelink access to the SSID
14. Tap on Allow for eWelink access to the smartphone's location
15. Wait while the info is transferred to the Sonoff 4CH Pro Rev2
16. Change the Device name to what you want to name it
17. Tap on Complete
18. Tap on OK
19. After the Sonoff has attached to your Wifi, you will be returned
    to the eWelink app screen
20. Tap on each of the outlet buttons and watch the Sonoff 4CH cycle through each

To use the Sonoff 4CH Pro Rev2 as a garage door

controller, we need to make a couple of changes. The relay inputs that you want to control a garage door will need to set to inching mode with a duration of 1 second.

1. Disconnect the Sonoff 4CH Pro Rev2 from power
2. Turn the 4Ch on its face
3. Remove all 4 screws
4. Remove the circuit board from the housing
5. Move the S6 switch to the other position (this changes the 4CH to inching mode)
6. Move all the dip switches on the K6 block to On or 1 (this sets the 1 sec delay)
7. Put the board back in the case
8. Install all 4 screws back in their holes and tighten them

At this point, you can take a pair of wires from each input on the 4CH Rev 2 and connect them to the screws on the back of each garage door controller.

Test each garage door as you have them connected to the 4CH Rev 2.

Video

https://youtu.be/uwFmfC0iJiY

Sonoff 4 CH Pro Smart Switch v2
http://ronaldnutter.com/s4chprov2

6-Foot 16/2 SPT-2 3-Outlet Indoor Cord, Brown
http://ronaldnutter.com/6ftbrwn

Klein Tools 11054 Wire Stripper and Cutter
http://ronaldnutter.com/kt11054

Tripp Lite 7 Outlet Surge Protector Power Strip
http://ronaldnutter.com/tlv7sp

Controlling Sonoff 4CH Pro Rev2 with Alexa

These instructions assume that you have already completed the basic setup of the Sonoff 4CH Pro Rev2 as outline in the previous chapter.

To enabled the eWelink skill, follow these steps -

1. Open the Alexa app on your SmartPhone/Tablet
2. Tap on the menu in the upper left hand corner of the app
3. Tap on Skills
4. Enter EweLink in the search box
5. Tap on the magnifying glass to begin the search
6. Tap on eWeLink Smart Home Fan
7. Tap on Enable to enable the skill
8. Enter your eWeLink account information
9. Tap on Login
10. Go back to the Skills page
11. Tap on the Discover Devices button

Edit the name that Alexa discovers to something a little more user friendly

If you have already enabled the eWelink skill, do one of the following -

1. Ask Alexa to Scan for New devices
2. Select the Scan for New Devices option in the Alexa app or Portal

To open or close a Garage door, ask Alex to turn <input_name) on

Using the word Open or Close doesn't work with the Alexa integration at this point. If you want to use something like IFTTT to get this keyword working for you.

Video
https://youtu.be/nEW1otL2h-M

Sonoff 4 CH Pro Smart Switch v2
http://ronaldnutter.com/s4chprov2

6-Foot 16/2 SPT-2 3-Outlet Indoor Cord, Brown
http://ronaldnutter.com/6ftbrwn

Klein Tools 11054 Wire Stripper and Cutter
http://ronaldnutter.com/kt11054

Tripp Lite 7 Outlet Surge Protector Power Strip
http://ronaldnutter.com/tlv7sp

Controlling Sonoff 4CH Pro Rev2 w/ Google Asst.

These instructions assume that you have already completed the basic setup of the Sonoff 4CH Pro Rev2 as outline in the previous chapter.

1. If you havent already installed the Google Assistant, Google Home and Ewelink apps, do so now
2. If you have already installed both apps, make sure you have the latest versions installed
3. Go to the Ewelink app on your smartphone / tablet
4. Tap on the menu in the upper left hand corner
5. Go to Connected Home
6. Click on the Plus symbol in the lower right hand corner of the screen
7. Tap on "Smart We Link"
8. Enter your eWeLink credentials on the next screen
9. Tap on Login
10. You will see a message about linking your account
11. When that disappears, you should be able to turn the Sonoff off and on

If you have already enabled the eWelink skill, do one of the following -

12. Ask Alexa to Scan for New devices
13. Select the Scan for New Devices option in the Alexa app or Portal

If you have already linked eWelink to your Google Home account, you will need unlink and relink your account for it to know about the Sonoff 4CH Pro Rev2.

To open or close a Garage door, ask Google to turn <input_name) on

Using the word Open or Close doesn't work with Google Assistant integration at this point. If you want to use something like IFTTT to get this keyword working for you.

Video

https://youtu.be/DM0mb1_iTbA

Sonoff 4 CH Pro Smart Switch v2

http://ronaldnutter.com/s4chprov2

6-Foot 16/2 SPT-2 3-Outlet Indoor Cord, Brown

http://ronaldnutter.com/6ftbrwn

Klein Tools 11054 Wire Stripper and Cutter
http://ronaldnutter.com/kt11054

Tripp Lite 7 Outlet Surge Protector Power Strip
http://ronaldnutter.com/tlv7sp

Controlling Sonoff 4CH Pro Rev2 with Cortana

These instructions assume that you have already completed the basic setup of the Sonoff 4CH Pro Rev2 as outline in the previous chapter. Since there isnt direct integration between Sonoff and Cortana, you can use a 3rd party, IFTTT (IF This Then That) to bridge between the two.

1. Create an account on IFTTT
   a. Go to www.ifttt.com
   b. Create an account on www.ifttt.com
   c. Download ifttt app for your smartphone
   d. Run the ifttt app
   e. Enter your login credentials
2. Connect Cortana to IFTTT
   a. Tap on the My Applets icon on the lower right hand corner of the screen
   b. Tap on Cortana or search for it and then tap on Cortana
   c. Tap on Connect
   d. Enter your Microsoft Cortana credentials
3. Connect Sonoff to IFTTT
   a. Tap on the My Applets icon on the lower right hand corner of the screen
   b. Tap on eWelink or search for it and then tap on eWelink
   c. Tap on Connect
   d. Enter your eWelink credentials
4. Create Applet on IFTTT to Open on Sonoff 4CH input/door
   a. Return to main IFTTT screen on your smartphone
   b. Tap on the My Applets icon on the lower right hand corner of the screen
   c. Tap on the plus symbol on the upper right hand corner of the screen

d. Tap + beside this on the New Applet screen
e. Type cortana on the search line
f. Tap on Cortana
g. Tap on Say a Specific Phrase
h. Tap on What do you want to say
i. Type open door_name door
j. Tap on Done
k. Tap on What do you want Cortana to say
l. Type opening trailer door
m. Tap on Done
n. Tap on check mark on the upper right hand corner of the screen
o. Tap on plus sign beside that
p. Tap on Services
q. Type eWelink
r. Tap on eWelink icon
s. Tap on Turn 4 channel plug on or off
t. Select the Sonoff device you want to control
u. Select the channel to control
v. Select on from the turn on or off option
w. Tap on Checkmark in upper right hand corner of the screen
x. Temporarily enable notification when this applet runs to help in initial debugging
y. Tap on Finish

5. Using either the Cortana app on your smartphone or the Harmon Kardon Invoke, try opening one of the garage doors

To open or close a Garage door, ask Cortana to open <input_name)

You have the option of either using Open (Input Name) to open and close the door or create an additional IFTTT trigger named Close (Input Name).

Video

https://youtu.be/fv3lbzZfvRs

Sonoff 4 CH Pro Smart Switch v2

http://ronaldnutter.com/s4chprov2

6-Foot 16/2 SPT-2 3-Outlet Indoor Cord, Brown

http://ronaldnutter.com/6ftbrwn

Klein Tools 11054 Wire Stripper and Cutter
http://ronaldnutter.com/kt11054

Tripp Lite 7 Outlet Surge Protector Power Strip
http://ronaldnutter.com/tlv7sp

Protecting the Opener and Controller

Installing the Garage Door Opener and Smart Controller is just one piece of the puzzle. Depending on how clean your power is can have a lot to do with prevent premature failure of one or both items. How clean the power is refers to how the voltage level spikes beyond normal levels or how much the power drops to levels way below what is considered normal operating levels. You can't protect against all power issues but plugging a device known as a surge protector in to the outlet you would normally use for the opener and the controller and then plugging those same device into the surge protector is one way of protecting your investment. When there is a power outage, the incoming rush of power can do more damage than you might expect.

Tripp Lite 3 Outlet Surge Protector
http://ronaldnutter.com/tl3osp

The garage door openers that I have seem usually have some type of electrical outlet installed near the opener. A periodic check of the status lights on this type of device should give you an early indication that it has either failed and/or needs to be replaced. Don't let the lights be the only thing you check. Surge Protectors don't last forever. Depending on how many power surges you experience over time, the amount of protection will gradually lessen. There are several schools of thought as to when to replace surge protectors. I have heard replacing the devices at intervals of 3 or 4 years. I would check this with the tech support o the company whose surge protectors you do purchase to see what they recommend.

Moisture Sensor

Samsung SmartThings Sensor
http://ronaldnutter.com/sstsw

Another area that I was looking for a Smart Home solution to help address was an intermittent presence of water I initially in my basement. The cause was found when I had the foundation repaired and was told that the root cause was no drain tile was put in when the house had been originally built. I was looking for something that would be able to tell me when even a small amount of water was present on the floor. This could be related to an occurrence of the leak or the failure of a sump pump to keep water out of the basement.

Since the likelihood of water being present on the floor could also be related to a prolonged power loss that could prevent the sump pumps from doing their job, to be fully protected this would mean also having a UPS protecting both the internet connection modem and the Samsung SmartThings hub so that alerts could continue to go out. Ideally having a battery backed up sump pump would be a prudent option to consider, the costs I have seen for this option make it almost cost-prohibitive when compared to the low number of power outages I had been experiencing.

Smoke Alarm

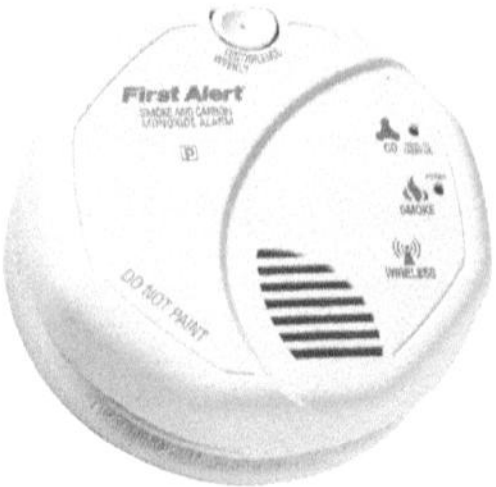

First Alert 2-in-1 Z-Wave Smoke Detector
http://ronaldnutter.com/fazsd

This is an example of a 3rd party device that requires the presence of a Z-Wave hub in order to function. This is where the Samsung SmartThings hub comes into play. This Smoke Alarm from a well known company, First Alert, that gives you a Smoke Alarm and a Carbon Monoxide detector in one package. Once you have this running in the SmartThings app, you can look at setting up additional steps such as automatically alerting when smoke or Carbon Monoxide is detected.

Samsung SmartThings Outlet
http://ronaldnutter.com/ssmout

This is an item that allows you to control items that might not be controlled through other options. The description for this item indicates that it is capable of controlling items such as coffee machines, Crock Pots, Portable Heaters, Fans and Air Conditioners. There are two models of this device, the square one you see here is the original model, a round one is the latest version of this.

The specification for this device indicates that it is capable of support a device running on 120 volts ac with a current capacity not to exceed 12 Amps. The number of Amps mentioned is an important thing to be mindful of. Some devices temporarily pull in excess of that when first starting up. If that is the case, then you cannot safely use this device. The documentation is very specific about that.

This is one of those items where it pays to read the small print. It may not be possible to automate the process of turning something on by a predefined schedule. Look at it this way - what might happen if you left a coffee pot on and little to no water was in the pot. For example, I have an Oreck Air Purifier connected to a Samsung Smartthings Outlet. I can turn it on via the Smartthings app but am able to turn it off via voice control.

Thermostat

This is an area that I wanted to get to a little quicker than I did. I thought that it would be just a matter of disconnecting one thermostat and connecting another one. Depending on how recently you have replaced your HVAC system, it may be easier for some than for others.

There are a series questions you need to get answers for to see if you can implement a smart thermostat or need to get some things changed or upgraded first. The first question is to look at the thermostat, inside the cover or with the thermostat removed from the wall to see if 110V is visible anywhere. If it is, then may have to look at a newer HVAC system or talk with your service company to see if there an upgrade that will take you to a low voltage system that operates in the 24v range.

The next question is, with the thermostat removed from the wall, if you have thick wires coming out of it that probably have what is called a wire nut that is connecting the wires coming out of the thermostat to the cable coming out of the wall. If that is the case, then you will have to look at some changes to your existing HVAC system in order to be able to use a newer thermostat.

The next, and possibly last question, is to see if you have a wire labeled "C" or one connected to a terminal labeled "C". If you have reached this point without any issues, congratulations you are at a point where you can start looking for your new smart thermostat. If you don't have a wire labeled "C", keep reading as you still may be able to implement something new.

See if you have an extra wire that isn't connected to a connection on the wall plate the thermostat mounts on. If you do, you may be able substitute this wire for the "C" wire that isn't connected. You will want to find the other end of the cable connected to a circuit board at the furnace to see if it is un-connected at that end as well.

If you didn't have a un-used wire, see if you have a wire connected to a terminal labeled "G". If there is such a wire, then you may be able to use it in place of a "C" wire. Consult your current HVAC vendor to see what your options are. This extra wire is needed to power the Wifi component of the smart thermostat.

CHAPTER SIX

Smart Home in your Auto

Android Auto

Yes, you read the title of this chapter right, Android Auto. Your first statement will be what does this have to do with Home Automation. My answer is more than you might think. There are several components in common between getting your home to work for you, your cell phone and your vehicle. If you have just started to say “My vehicle doesn’t have Android Auto”, don’t let that stop you. I will show you how you can leverage your cell phone even more than as an interface to control your Smart Home.

What you need to use Android Auto

1. An Android based phone

If you have an iPhone, I am not aware of any plans to bring the Android Auto app to the Apple platform. Those who know me have made a little bit of fun of my having both an iPhone and an Android phone for several years. I was one of the testers of the Google Project Fi cellular service. If you are haven't heard of Project Fi, let me take a few minutes to introduce you. The biggest attraction for me is that Project Fi released me of the shackles of being tied to a particular carrier. Google has purchased minutes for a collection of carriers. Your phone looks for the best tower signal from the carriers that are current a part of the Project Fi network. Only once, in a very remote area, have I experience any connectivity issues with finding a good connection to a nearby tower. The good news here is that you don't have to have a top of the line phone to use Android Auto, so you can start with what you can afford now and then upgrade to something better once you are comfortable with what AA can do for you.

2. Power cord and adapter for your vehicle

If you have done much of anything with whatever cell phone you have, the more you are doing at any one time will use more of your battery capacity. I found that when running any type of GPS or traffic app, having the phone connected to the power in the vehicle will keep the battery charged as much as possible and still leave you a usable cell phone. In my EDCC (Every Day Carry Kit), I keep an power cable for my Android phone and a USB Cigarette Lighter plug. With this, I am able to move to a vehicle other than mine with a minimum of hassle.

3. Android Auto app

This is the foundation piece of having Android Auto working in your vehicle. Download the app and get it started up. It should pick up your Google login and then will ask you a series of questions asking for the permission it needs to do it's job. Android Auto will act as your hands free interface to driving to/from your destination, making/receiving calls and listing your your favorite music/podcast. Your journey will start with saying "Hey Google".

4. Waze app

This app is one that I have used for a while long before I heard about Android Auto. I have used a variety of GPS systems over the past few years. In that time, I found consistently better value and results with Waze. There was a lot of concern when it was announced the Google bought Waze. To Google's credit, Waze has continue to operate on its own. Waze is a crowd source traffic monitoring/reporting system. In the past, I would usually find out about an accident or traffic problem when I got close to it. With Waze, I not only get advance warning of such problems but also get a heads up on broken down vehicles on the side of the road or other road conditions that allow me to avoid becoming involved in a problem.

In order to get full use from Waze, you will need to create a login account. After you have done that, go ahead and log into Waze. I would suggest going ahead and creating two favorites, one for your work address and one for home. I did this when I first started using Waze before Android Auto was on the scene and it made things easier to work with.

Part of my reason for starting with just the basics is that I want to get a baseline idea of the bandwidth that using just Waze and Android Auto will have on my monthly data usage. That and following the principle of "Keep It Simple Stupid" will make sure that I have a good handle on what just the basics will do and the commands that I need to use before I start turning on all the bells and whistles.

As you read more about Android Auto, you will see that it has a lot of potential and a lot to offer. What I have discussed so far is that I am starting with and what I would suggest that you do as well. Trying to do everything at once will be like trying to drink from a firehouse connected to a fire hydrant that has been opened all the way.

Smart Home in Your Auto

Before you say "I don't have a Smart Home in my vehicle", you do. More and more cars in recent years have had an increasing amount of automation built it. That is the problem, it is built in and disconnected from what you have in your home.

There are some options where you can control some or all of your smart home from your vehicle, otherwise known as your home on the road. Some Smart Home options have the ability of controlling your devices when out of the house but not all Smart Home platforms support this. Those that do, such as Apple's HomeKit, require that you have an additional device such as a Apple TV or iPad to serve as a base station with a secure connection to the Apple cloud in order to accomplish that.

As with anything that you are trying to accomplish, where there is a will, there is a way. If you have one or more Smart Home platforms or devices that don't support remote connectivity, you may have a couple of options. Option one requires some type of desktop or laptop for you to remote into using something like TeamViewer or other remote control application. The issue here is that if you aren't connected, you may not know there is a problem until it may be too late. This also requires that you dedicate a computer to be used for this purpose. For those smart phone platforms or devices that don't have an app that you can manage/ monitor those devices from a laptop or desktop, you may have to look at using an old smartphone or tablet as a point to take remote control on. Either way isn't without its pros and cons. This is where the two areas that follow, Android Auto and Alexa in the Car are options worth considering.

Alexa in the Car

The first time I thought about this possibility involved a mobile hotspot, an Amazon Echo Dot, a power inverter and/or a cell phone. It didn't take long to rule this out for several reasons. There were too many pieces involved and too many areas where there could be a problem. I knew there had to be a simpler way.

Roav Viva

http://ronaldnutter.com/viva

Just when I thought there weren't going to be a viable option, wouldn't you know that just after CES (Consumer Electronics Show) 2018 that Anker released the Roav Viva. Basically it is a fraction of the size of an Amazon Echo Dot but in a form factor that easily fits in most cars. The Roav Viva isn't just another pretty face. Besides giving you a way to use Alexa in your home away from home (aka your vehicle), it also gives you two USB ports equipped with PowerIQ charging technology. While this may not sound like much, let me put it another way, it has the potential for reducing the charge time require for a device by as much as 1/3 of what a regular charger would need.

Controlling Your Home from the Car

This isn't as far fetched as you think. With the Roav VIVA, you are already talking to Amazon Alexa. If you have Alexa talking to other devices in your home, there isn't any reason that you should be able to control anything that Alexa is already controlling from your home but now from your vehicle. Depending on some of the third party integrations or "skills to the the Alexa terminology, you might have to separately authorize the Roav VIVA to some of those skills but I would expect that to be the exception to the norm.

Getting Started with the Roav VIVA

Before or when you get your own Roav Viva, get the Roav Viva app that is available for both iOS and Android platforms. Once you have downloaded the app for your phone, follow these steps to get started -

1. Tap on the Roav VIVA app to start it

2. Tap OK to allow Roav VIVA to access and sync your contact list locally so you can make hands free calls by using voice commands

3. Tap OK to allow Roav VIVA to send you Notifications

4. Tap "Always Allow" when the Allow Roav VIVA to access your location. Hopefully Roav will set up a timeout option where if you aren't moving or in the vehicle, that it wont try to access location services. When using an iPhone, the easiest way I have found to deal with this and not have to get into the menus to disable location services is to double tap the home button on an iPhone and when I have the Roav VIVA app screen showing, swipe it off the screen.

5. When the End User License Agreement (hence known as EULA), tap on Accept to continue

6. You will now be at a screen where you will have the option of login in with your Facebook, Google or Email account. I prefer to use my email account as I generally prefer not to use my credentials for one service to login to another. Tap on Create Account to continue.

7. Tap on Email to enter your email address

8. Tap on Password to enter the password you want to use for this login

9. Tap on Create Account

10. Plug your Roav Viva into the cigarette light outlet in your vehicle

11. Follow the instructions on the screen to being the pairing process between the App, your phone and the Roav VIVA.

Video

https://youtu.be/Bdd_Rpnf65Y

Bluetooth Pairing

I found that the iPhone, at least in my case, has been a little slower to pair. Some of this could be unique to my situation. I will continue to watch this. My iPhone was on release 11.2.5 when I first started using the Roav VIVA. Just to make sure that I hadn't overlooked any possible option. I checked for an available update and found that 11.2.6 was available. That update has been applied and I will report on what issues have changed as a result of that update after a few days of working with the newer firmware. I have had an issue with another bluetooth device that had issues with 11.2.5, so I wasn't surprised when I had an issue with pairing the VIVA.

Power IQ Charging Ports

Consider me a little jaded. I initially thought that this term might be a little bit or marketechture. In just the first drive, I was surprised with what I found. For the most part, my daily driver phone is the iPhone 6 Plus. During my drive, I have the follow apps open and running - Waze, Apple's Podcast app and the app for my Escort radar detector. Two of the 3 apps use GPS heavily. If I start out with the phone at 100% charge, by the time I get to work (about 40 minutes), the charge level (with the phone plugged into the vehicle's USB port) will drop to about 92% or lower.

When I got into the truck for the drive home, the iPhone was showing a 78% battery level. With the same three apps open and a 40 minute drive completed, the phone was showing a 92% battery level. I would have been happy with keeping the battery at a 78% level, to have a higher level of charge was a pleasant surprise.

Battle of the Audio apps

As I mentioned earlier, my normal routine during the drive to/from work is to listen to a collection of podcasts that I listen to regularly. I had the Apple Podcast app running and wanted to listen to some of my music collection. When I gave the command to listen to my music, Alexa immediately said that another app was using the audio channel and I needed to stop it. That was followed at almost the same time as a popup window near the top of the screen that asked about giving control to the music I was trying to play. Nice touch that I preferred over inadvertently stopping what I was listening to.

Looking for commands VIVA would respond to

Ok, I will admit that I didn't read much of what the Roav VIVA came with. In part, that was my attempt to see what commands I could find that it would respond to. One of the commands I tried was "Alexa, what are my GPS coordinates ?". To my surprise, the VIVA responded back with "Your coordinates for zip code XXXX is " and what followed were the coordinates that it was able to identify.

At different points on the trip home, I asked "Alexa, what are the current road conditions?". I got a response that indicated the current town I was in and the approximate travel time. At the time I had tried the commands, the travel conditions were changing, so it is possible that the information that VIVA was consulting wasn't getting update either quickly enough or in enough detail to that info. I do live in a metropolitan area where road conditions are reported by two different state highway departments. That may not be the case in all areas.

Roav VIVA, Alexa Enabled
2-Port USB Car Charger
http://ronaldnutter.com/viva

Update on IOS 11.2.6

On the morning drive, I didn't have the problem with the app disconnecting from the VIVA as I had seen when on 11.2.5. I noticed that the pairing process after the phone has been powered up and the VIVA has completed it startup process is now more consistent. The VIVA app now automatically starts up on the iPhone and has connected VIVA device with no intervention from me. On the afternoon drive, I was prompted by the VIVA app to start the pairing process one time and after a stop, it automatically handled the process without intervention on my part.

Using Waze instead of Google Maps

One of the things that attracted me to Waze is it's crowdsourcing of traffic information. Even though I live in an area where there are two state highway departments providing reports, there is enough delay that the information, especially during rush hour traffic, sometimes isnt the most up to date. The first thing I needed to do is change what mapping program that Roav VIVA runs when I give a command of where I want to go. You do this by going into the Roav VIVA app and clicking on the gear icon on the right hand side of the page. Go down the page until you see the Navigation option. Tap on it to make the change you need. You will have three options - Google Maps, Waze and Built-in Maps. Tap on Waze to select it. You should be returned to the VIVA apps settings screen you started from.

In using Waze with VIVA, I notice that I still have to tap the Go button in Waze to start the traffic process. This is probably a safeguard on Waze's part. That is something that I will be running down to see if there is a better way to deal with the handoff between the apps.

Podcasts with Alexa and TuneIn

Here are commands that I have tried -

Alexa, Play "Name of podcast" podcast - This will play the latest version of the podcast (if available via TuneIn)

If you podcast isnt available, reach out to the host and have them fill out this form - https://help.tunein.com/customer/portal/emails/new?ticket[labels_new]=podcast&t=764630 to get it listed

Alexa, Stop - stops playing podcast
Alexa, Play the previous episode

Volume settings

- "Alexa, louder." Turns the volume up a step.
- "Alexa, softer." Turns the volume down a step.
- "Alexa, set volume to 5." Sets the volume to (0 - 10).

- "Alexa, mute." Mutes sound.
- "Alexa, unmute." Unmutes sound.

Wake word vs hitting the button

During my daily commute, I usually listen to a collection of podcasts what I keep up with. I don't play them loud by any means. I find that VIVA doesn't always "hear" me. Probably because it is trying to "listen" to what is being said on the podcast being played. Cant fault the VIVA on that. So far, the solution is to turn down the stereo or tap the button on the front of the VIVA.

On the drive home, I played music instead of listening to the podcasts that I usually do. It was easier to get VIVA's attention during the music. Depending on what type of music was playing, sometimes I did have to use VIVA's button to work with Alexa. This should change over time as we learn to work with each other.

Hand-Free Cellphone use

I had the opportunity today to place a hands-free call with VIVA. After getting VIVA's attention, all I had to say was "Call" followed by the name of the individual I wanted to call. The person I needed to talk to had three different phone numbers. Viva quickly displayed the three numbers and gave me a quick way to indicate which of the numbers I wanted to use.

Video

https://youtu.be/a9LDd8tlYQ8

CHAPTER SEVEN

SmartHome for Travelers

SmartHome for Travelers

When you get used to having your own Smarthome, it would be nice to have at least a part of it when you have to travel. Dealing with networks at hotels and other venues can be a challenge. Above anything else you need to protect your own equipment. The AR-750 travel router that the following chapters are built around is one of those options. There are and will be other travel routers that you can use.

One specification that I would suggest you consider using is a travel router that has both 2.4Ghz and 5Ghz wifi bands. Most hotels and venues will use 2.4 because it is cheaper to implement and has a wider coverage area than 5Ghz does (in general terms). In this type of deployment, you would use the 2.4Ghz portion of the travel router to connect to the venue's network and then 5Ghz to connect to your equipment. That is just one way that you can connect to a venue's network. You have to take every reasonable option possible since you don't maintain the venue's network or make sure that certain precautions such as keeping the latest firmware and patches installed are being done per OEM recommendations.

Video

https://youtu.be/oWZL7k5dlJ0

Equipment Options

Having a Travel Router should be the foundation of protecting any type of technology that you have with you when traveling. With more and more hotels having open Wifi which means no encryption or keeping someone from watching what you are doing, having a way to protect yourself becomes even more important. The AR750 offers several advantages that make it worth looking at. You have a well known firewall protecting your laptop or any other technology with you. Currently, the AR750 is using the OpenWRT firewall. At some point, DD-WRT may become an option as well.

Just as important as a good firewall is being able to protect your data from prying eyes near you. This is done by including the ability to have the OpenVPN technology as a part of the firewall. Especially when you have to use an open/unencrypted wireless connection, using a VPN as a further level of protection is nice to be able to have. Even when on a wired connection, this is worth having because if you didn't put the network in that you are using, how much are you able to trust how safe your traffic is from prying eyes ?

This is one of those areas where you get what you pay for. Until you see how much you may be traveling, there are a variety of free VPN services that you can use to get used to the process of using a VPN. Even when you are using services that have encryption built in, having a VPN adds an additional layer of protection. Setting up the OpenVPN client couldn't be easier. In my case, I went to vpnbook.com. You select the VPN server you want to connect to and download the .zip file containing all the files you need. The VpnBook.com website also gives you the user credentials you will need when installing the config files on your AR750. After a minor configuration change, you can enable the VPN by the press of one button. You can't make it any easier than that. For each VPN server you select from VpnBook, you have four different ways to connect so that you should be able to get a secure connection in just about any situation.

Video

https://youtu.be/oWZL7k5dlJ0

Travel Router

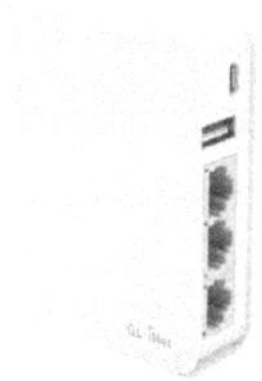

GL.iNet GL-AR750 Travel AC Router

http://ronaldnutter.com/glar750

Shockproof Carring Case

http://ronaldnutter.com/ar750case

Streaming Media Receiver

This is the next step in having a portable smart home. Even when first starting out, this is probably one of the first things that you will have. Before you say, this wont be that useful when traveling, if you have any degree of internet speed at the hotel you are in, you will at least have some options of what to watch instead of the potentially limited offering or having to resort to PayPerView.

You have several options for a portable streaming media device. I started with this one because it also has the Alexa Voice Remote. That should save you having to go all over the screen to select something you want to watch from your Amazon Prime benefits or other services such as SlingTV. Since more TV's in hotels have at least one HDMI connection, a portable streaming device has the potential for giving you a little slice of what you have at home. Another benefit is that when you add a portable Plex media server, having a streaming media device will give you the ability to run the necessary client app to be able to see what you have on your Plex server.

Amazon Fire TV Stick
http://ronaldnutter.com/ftvs

Plex Server

The Plex media server gives you an option of watching movies from your personal collection, pictures from that last vacation or the music you like to play at various times. You have several options as to what to use for a Plex server. You might be able to use the laptop that you are traveling with but if it is locked down like some corporate issued laptops are, you might need to look at options such as a Mac Mini, an Intel NUC, small laptop or even a Raspberry Pi Single board computer. When using a device like the Raspberry Pi, you will want to consider an external USB hard drive to make it easier of having your movies or music with you.

Voice Control

I am listing two different options here. The Dash Wand is making an appearance because it is a light weight compact device. One of the few things it can't do is to place music. I have one handy because it is battery powered and can be in your pocket without a lot of thought. While you can have the Alexa client on your Android smartphone, you may be using the phone for something else that would make it temporarily unavailable to use as an Alexa device. Hopefully at some point, that functionality will be available on Apple devices as well.

Amazon Echo Dot
http://ronaldnutter.com/aedot

Amazon Dash Wand With Alexa

https://www.amazon.com/Amazon-Dash-Wand-With-Alexa

LIFX Mini White Smart LED Bulb

While this might not be one of the first items you would add to your traveling SmartHome, it is worth considering. How may times have you entered a hotel room in the evening and have had to feel your way around the walls to find the light switch ? Combining this and the Amazon Echo (or potentially Google Home) and you have the ability to turn on a light before you even enter the room.

http://ronaldnutter.com/lifxmdd

Implementation Levels

As with the smart home tech you have in your house, there are multiple levels or ways that you can implement a similar level of technology for when you have to travel. How you implement the smart tech you will be traveling with will depend on how much you travel and for the length of the trips. The more you travel and for long periods of time will help justify how sophisticated your traveling smart home setup may be.

Level 1

This is where you will create the foundation that the rest of your traveling smart home will be based on. At a minimum, you will want a firewall that can protect the equipment you are traveling with and a network (either wired or wireless) that you are connecting to. Having some type of streaming media receiver (probably in a stick form factor to keep it small) will at least give you some options to what might be a sparse number of local and/or satellite channels to choose from.

My goal with having a traveling smart home is to have it as compact as possible. In the setup I am currently using, I have the travel router packed in its own case that houses the router, the power supply and the minimum cables required to get it working. The case I chose may be different from what you might go with. You can go from a soft case that I use to something a little more rugged like a Pelican brand case. Don't try to buy a case right a way, you can use something like a plastic baggy to get started with. When you go for a more substantial case, try arranging what you want in a particular case to get an idea as to what the ideal internal dimension will be for the case that will work best for you.

Level 2

With Level 1, you get the basics of a smart home that will give you the bare minimums of what you had at home. The more versatile you can have the Level 1 part of your traveling smart home in terms of connectivity and cable that can be used, it will help give you some options at Level 2. This is where a Plex server (albeit it a smaller one) can give you the ability to playback some of your favorite TV shows or movies when you want to see them.

At this level, you will probably have two components, the Plex server and some type of external USB hard drive. You may find that this level may take a little more TLC and/or tweaking to get things running. If your trips are mostly overnight trips, then a Level 2 version of a traveling smart home may be a little project to keep you busy when you don't see anything you want to watch or just need something different to do.

There are three ways you can implement a Plex server. For those who are looking for a way to let their inner geek out, you have several options to work with here. If the laptop you are traveling with hasn't been locked down by Corporate IT, you can install the Plex server on it and attach a portable USB hard drive containing the Movies and Television shows that you want to watch. The advantage here is that you are up an running with a minimum amount of hassle.

If that isn't an option for you, you can look at either a Raspberry Pi or a Mac Mini. In terms of form factors, the Raspberry Pi will be the smallest and least expensive option. The limitation is that you will only be able to treat to a single TV. Since the Raspberry Pi is what is known as a Single Board Computer (SBC), there is a limit to what it can do in terms of processing power. You can get a starter kit with everything you need to get started for around $50. One Plex distribution that I have been working with is called PlexPi (available at PlexPi.com). The ideal thing with this particular PlexPi implementation is that it is as close to plug and play as you can hope to find. Since RaspberryPi usually runs linux, you might think that you will need to know a lot about linux. That isn't the case with this implementation. You download the image file and burn it to a miniSD card. Put it in the RaspberryPi, connect it to your TV and power supply and you are up and running in a few minutes.

One thing to remember about the Raspberry Pi is that with it being a single boar computer, the heat that it generates has to go somewhere. One kit that I have used includes user attachable heat sinks that can be mounted in just seconds. When you are using the Raspberry Pi to handle video processing, there may be a limit to the amount of heat that can be dissipated in given period of time. When heat gets to a certain level, the CPU will start slowing down in response to temperature. There are some cases that offer additional levels of cooling and/or heat dissipation to help the chips that are getting hot from loosing too much of there cool.

If you have a little more room and can spend a little more money, a Mac Mini might be worth looking at. It is a little more expensive but has more resources in terms of CPU and RAM to handle some tasks you may need such as transcoding. If that isn't a term you are familiar with, it means that if the movie you are trying to play is mean for a TV but you are playing it on a smartphone or tablet, the Plex server will have to convert or "transcode" the file into a format that can be streamed to the device you are watching the program on.

Level 3

This is the next level to your traveling smart home. How many times have you opened the door to your hotel room and had to feel around the wall to find the light switch ?

With my luck there was more than one light switch and I always turned on the wrong light. With this level of smart home, you have the choice of either turning on the light with your choice of Amazon Echo, Google Assistant or Microsoft Cortana.

Here is where you may have to make a choice on how to implement it. For a portable traveling solution, the Amazon Echo Dot or the Google Assistant Mini will be the easiest two form factors to work with. The 3rd part hardware options for Cortana are a little more limited. Microsoft has gone the software client route, so it is up to 3rd party manufacturers such as JBL and Harman Kardon to come up with an option if there enough demand in the marketplace to support it.

In a previous chapter, I mentioned the Roav VIVA. While intended for your vehicle, there is a way that you can use it as a part of your traveling smart home. For between $12 and $20, you can find an AC Power Supply that gives you the 12 volt DC needed by the Roav to work. In this case, there is one more item you will need. When you are using the Roav Viva in your vehicle, it uses your smartphone as an ad-hoc internet hotspot to be able to get what it needs out of the Amazon cloud. I have looked for a bluetooth to ethernet adapter to get around having to have a smartphone but so far haven't been able to find that type of product.

You will need to have some type of smart led bulb to use. Elsewhere in this book, you will see discussed several options. Something like the LIFX bulbs may be the easiest to work with as they have built in wifi (most like 2.4ghz only) to connect to your traveling network. Some options like the CREE connected LED bulb will need another device known as a gateway or hub that incorporate a protocol known as Zigbee or Zwave. These protocols allow for a device that doesn't have support for wifi to talk over a network to do your bidding.

As with your smart home tech you have at home, there

are a variety of options that you can take traveling with you as well. The light bulb we discussed earlier can be something as simple as the bulb and a socket that plugs into the wall. That would give you at least a night light or minimal light to have some light in the room until you can get other lights turned on. There are also a variety of plugs you can use to turn lights or other devices on or off as your situation dictates. This is where having a checklist with your traveling smart home will help ensure that you don't leave something behind.

Level 4

This is a level that I am working towards. The more I thought about a traveling smart home is having some options to having some type of "smart" alarm system. For infrequent trips that are only a night or two, this probably wouldn't make a lot of sense. On the other hand, having some of type of ability to record when the room was entered or a video record of what happened while you were gone or motion was detected could go a long way to getting something back that was missing because you will have a record of the event as it happened.

Things to remember

The main thing that you will want to keep in mind is that everything you have in your traveling smart home will need to be portable, compact and not require any type of mounting on the wall. You should be able to pack and unpack your portable smart home within a few minutes depending on the number of items you have. While it isn't realistic to try to have all the comforts of home with you, you should be able to have enough with you to help minimize the pain of traveling away from home.

Bandwidth Requirements

Streaming Media has its own set of requirements. Each video resolutions has its own set of requirements. Please keep in mind that these are minimums and that depending on what codec was used or how the media is being streamed to you, more bandwidth than what is shown may be necessary for an acceptable viewing experience.

Resolution	Pixels	Minimum Speed
SD	480p	3 Mbps
HD	720p	5 Mbps
Full HD	1080p	10 Mbps

Depending on the speed of the network where you are at has available, you may or may not have consistent results. To give you an additional layer of protection beyond what you are using to stream video has available, using a VPN connection may be prudent option to consider. Keep in mind, that when you use a VPN, your actual network speed within the tunnel will probably be less than what you experienced without the tunnel running.

Wired or Wireless

This is another area in your traveling SmartHome that you may need to do some experimenting with. My general rule of thumb is to use wired whenever possible. Even with the improvements to wireless, when you are in a hotel environment, that will be the one thing that you little to no control over.

The rule of thumb I have found, especially when attending a conference event with several thousand people is that the available bandwidth may be hard to come by. That is where keeping everything using a wired connection may yield you the best results. When streaming a video from your portable Plex server, this should also help you get the best video to the TV as well. I have used wireless at times but have found that the results I get can be unpredictable.

If you do use wireless, I would suggest for the SSID that you setup for your own in-room use be named something that wont give others an idea of whose wireless it is or even worse, give clues to the password/passphrase for the SSID they will need to use. Please don't use something like Password or your date of birth for a password/passphrase. I will be the first to admit that s strong password/passphrase is a pain to type in but look at it this way, do you want to be an ISP for those in adjoining rooms ?

If the network bandwidth is really slow, there is an option that I would suggest to use only when there is no other choice. Depending on the type of smartphone you have and the cell carrier that you have, you might be able to look at something called tethering. The AR750 firewall that I mentioned earlier has support for this. You will use what looks like a charging cable to connect your smartphone to the AR750 (or whatever you are using for a firewall). IPhones seem a little more challenging to do tethering with when compared to my experience with Android smartphones. PdaNet is an app that I have used for several years that helps make the process a little easier to setup. There is a free and a paid option. The main difference is the paid option doesn't timeout and shutdown after a certain period of time.

GL-Inet AR750 Initial Setup

This section walks you through the setup of the AR750 to get you running quickly at a basic level.

1. Connect AR750 to WAN/Internet Connection
2. Connect your laptop to LAN port on AR750
3. Connect AR750 to AC power
4. Wait about to 2 to 3 minutes for the AR750 to fully power up
5. Open a web browser and go to http://192.168.8.1
6. At Choose Language Screen, click on English
7. At Choose Timezone Screen, click on the dropdown box to select your timezone
8. Click on Next
9. At the Set Your Password screen, enter the password you want to use in the designated fields
10. Click on Finish
11. Look at the bottom left of the screen, look for Firmware and note the version number
12. If this is less than 2.7, go to http://www.gl-inet.com/firmware/
13. Click on AR750
14. Click on v1
15. Click on lede-ar750-2.27.bin to download the file to your computer
16. Back on the main AR750 screen, click on the Firmware logo
17. Click on the Upload Firmware button on the upper right hand corner of the screen
18. Drag the file to the designated area of the screen
19. Once the file has been uploaded and verified, click on the Upgrade button
20. A dialog box will appear showing the upgrade process

21. The AR750 will reboot during the process

22. After the reboot, you will need to relogin to the AR750

The AR750 is now up and running on a basic configuration. You can use this for a wired connection from your laptop and to the hotel network

Video

https://youtu.be/dBAwGdIWr0E

GL-Inet AR750 Basic Setup

1. Connect AR750 to WAN/Internet Connection
2. Connect your laptop to LAN port on AR750
3. Connect AR750 to AC power
4. Wait about to 2 to 3 minutes for the AR750 to fully power up
5. Open a web browser and go to http://192.168.8.1
6. Open a separate tab on your web browser and go to https://www.vpnbook.com/freevpn
7. Download at least one certificate bundle
8. Keep the tape open as you will need other information from this site
9. Click on OpenVPN icon on the AR750 screen
10. Click on the Select Files text
11. Browse to where you downloaded the certificate bundle from vpnbook.com
12. Click on that file
13. Click on Open to upload the file
14. When the VPN Authentication screen appears, enter the username and password from the vpnbook.co screen you were at earlier
15. Click on the Submit button
16. Check the Enable box on the connection profile you want to use
17. Click on Apply to establish the VPN tunnel
18. The screen will show the update status at the tunnel is established
19. Allow several seconds for the process to start after you click the Apply button

To enable the slide switch on the side of the box to do this without using the GUI

1. Go to the main AR750 screen (http://192.168.8.1)
2. Click on the drop down arrow under Switch Button

Configuration

3. Select and Click on OpenVPN Toggle (On/Off)
4. Click on Apply

Note: The switch button on the side can only be assigned to one purpose. Any other use at the same time will require using the management gui. When using the slide switch to enable/disable the VPN connection, this will automatically select the last profile that you used. To check the status of the VPN tunnel requires that you go to the OpenVPN page of the AR750 screen. The AR750 is now up and running on a basic configuration. You can use this for a wired connection from your laptop and to the hotel network.

Video

https://youtu.be/P2lPqcdpvfo

GL-Inet AR750 Wireless Bridge Setup

This configuration is handy when you are in a location where you have one or more devices that need internet access and either they don't have the option to connect via a wireless connection.

1. Connect AR750 to WAN/Internet Connection
2. Connect your laptop to LAN port on AR750
3. Connect AR750 to AC power
4. Wait about to 2 to 3 minutes for the AR750 to fully power up
5. Open a web browser and go to http://192.168.8.1
6. Click on the Globe icon
7. Click on the Repeater tab
8. Select the SSID you want to connect to
9. Enter the passphrase for the SSID
10. Click on Submit
11. The AR750 will reboot
12. Once it connects to the configured SSID, you will see the ip address assigned on the Globe tab along with WISP
13. The Internet status screen may or may not show an update with more detail
14. Click on Home and Globe, you will then see the detailed information

Note: If you have an intermittent problem seeing 5 Ghz SSID's that other devices can see without a problem, try the following -

1. Reboot the AR750
2. Make sure you are on the latest firmware
3. If you are using Chrome as your browser, try changing to Firefox (that fixed the problem for me)

Video

https://youtu.be/DYQldgvgg0s

Setup of Amazon Fire Stick

1. Connect Power Adapter to the mini USB port on Amazon Fire stick
    a. If the TV you are going to use has a USB port, you may be able to use that instead of the power supply. If you get a message about insufficient power or compatibility message, you will have to use the power supply
2. Plug Amazon Fire Stick into an available HDMI port on the TV
3. Select the HDMI port on the TV
4. Install the AAA batteries in the remote for Amazon Fire Stick
5. Pair the remote to the your Fire Stick
6. Connect the Fire Stick to the Internet
7. Register your Fire Stick
8. Your Fire Stick is now ready to use

When using your Amazon Fire Stick as the media client for a Plex Server, you want experience better results when using a wired connection. Amazon has available an adapter which inserts in between the Fire Stick and the Fire Stick power supply. This is also worth considering to using when you have a wired network connection available where you are staying as well.

If you are limited to a wireless connection only, I would suggest using a 5Ghz based SSID as it will give you a better chance of a good streaming experience. This will help keep your streaming activity off of the hotel network. Especially if you are using the Amazon Fire Stick at home and as well as when traveling, you can either use the same SSID traveling or use a different SSID for when you travel (which would be my preference).

Video

https://youtu.be/saBAqeu4Lf0

Protecting Your Devices

The more devices you have working on your Smart Home, there are a few steps to thing about to make sure those devices stay running as long as possible. The easiest thing to do is to have put some type of protection in place to keep power surges from damaging the devices that are plugging into AC power. For some devices, such as the hubs or steaming media players, you might want to think about a UPS (short for Uninterruptible Power Supply). If you haven't heard of on before, a UPS has a battery in it and some electronics that cover the DC power coming out of the battery into AC power needed by the devices plugged into it.

By the time you get to this point of the book, you probably have a collection of devices doing everything from running your TV to automatically adjusting the thermostat based on your particular needs. What you have managed to put together is something called the Internet of Things. This term refers devices other than a computer that you don't login to but have some need of communication over a network to other device(s).

Surge Protectors

One item that you will probably want to purchase is a surge protector. This is something that will protect the various Home Automation devices that need AC power from being damaged from surges or spikes in AC power. Depending on the number and location of the home automation devices, you may need more than one of the surge protector outlet strips.

Tripp Lite 7 Outlet (6 Right Angle + 1 Transformer Outlet) Surge Protector Power Strip
http://ronaldnutter.com/tl7ospsp

APC 7-Outlet Surge Protector
http://ronaldnutter.com/apc7osp

Consider these devices something like an insurance policy. While to a guarantee that a device wont be damaged, it gives you some degree of protection from suffering damage. One thing to think about is that some of these surge protectors come with a protection guarantee that if their device doesn't stop the damage, they will cover some or all of the costs of replacing the damaged devices that were plugged into of the outlet strip that failed.

There are two schools of thought as to how often these surge protectors should be replaced. While there is usually at least one or two LED light on the protectors that tell you that protection is or isn't working, by the time those lights no longer work, you may have already had irreparable damage to one or more devices plugged into that protector. Replacing the surge protectors every two three years shouldn't be that costly since the average cost of the ones I have looked at runs the range of $10 to $15 each.

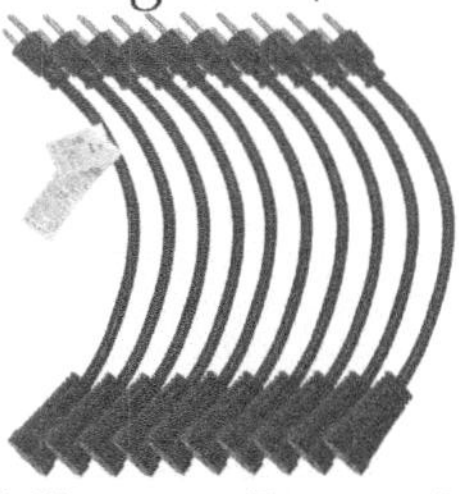

KMC Power Extension Cord

http://ronaldnutter.com/kmcpec

Another challenge you may run into is being able to use all of the outlets on the surge protectors. Not all devices use a simple plug for plugging into an AC outlet. Some devices use a small transformer, otherwise known as a wall wart. The problem is that a transformer, no matter how small may block some or all of a neighboring outlet. I have also seen this problem with a simple AC plug due to the power cable coming from the side of the plug instead of the top of the AC plug.

Using a small extension cord (usually 6″ to 12″) may be just what you need to be able to use all of the outlets on a surge protector strip without having to connect an additional outlet strip at the same location either from another outlet on the wall outlet or from a remaining outlet on the surge protector already plugged in to the last available outlet. Plugging in one surge protector into another one isn't something I like to do unless there is no other choice. Given the expected lower power requirement of a home automation device would be less than something like a television or small appliance, this probably isn't something to be overly concerned about, just something I prefer not to do.

Network Connectivity

I have waited until this part of the book to discuss a subject that may scare some people. Not everyone works in IT that has a smart home, so there may be some concepts that you may not have been exposed to before. For most, a network at home probably consists of a cable or dsl modem from your internet provider and some type of wireless access point. When you start working with a network that contains IOT (Internet of Things) devices, you may want to start thinking about some things a little differently.

The area of network security has become as important for the home as it is for companies. None of us may want to install all the updates in a timely manner on our windows laptops but with recent hacking attempts, do we have any other choice. This isn't just for windows but applies to all operating systems, even to those imbedded in the smart home technology that you are or are about to be using.

This is not something that you have to do immediately but it is something that you should start thinking about. There are a variety of ways that you can considering implementing additional levels of protection. One way you can implementation additional protection for your home network is to purchase a second router/firewall. With this firewall, you would implement a different SSID that only your smart home tech or IOT devices as they are more formally known would use. With this solution, if there was a problem, you would have the option of either disabling the device in question or disconnect the second router/firewall from its internet connection.

There are other options such as purchasing a more sophisticated wireless access point that would allow you to create multiple virtual networks. In the business work, this is known as a vlan. This means that you have to get a little more sophisticated network switch that that also has this type of support so that you could keep the network containing your smart home or IOT devices from talking with anything else in your home. This does introduce an additional step on your network where you will need to have some ability to temporarily move a laptop or other computer on to the IOT network or create some firewall rules that will allow a device from outside the IOT network to access the devices on the IOT network for configuration changes and or firmware updates.

This is not meant to scare you off but to help you understand that there are some additional things to think about. With the implementation on smart technology, we need to think about how to best protect the networks we have in homes. As the level of smart tech grows in our vehicles that is another area that will need to be looked at as well. Remember those stories how researches hacked their way into the networks on a certain brand of vehicle by using the built in cellular network connection in the vehicle ? We can only expect the manufacturers to do so much, ultimately we need to do some of this ourselves.

Choosing the Right Firewall

Some of you may or may not have heard about the FBI announcement/PSA for everyone to reboot their home router. Here is the interview I did for the local ABC affiliate where I live on this subject.

Video

https://youtu.be/aVFbB7TA8iU

There are several things that I said that didn't make the interview that was aired that are worth saying again -

1. Rebooting home router should be done at a bare minimum

2. See if the router manufacturer has a firmware update for your router.

   a. Download and apply it

   b. If a newer update isn't available, download the latest and apply it to your router. For some routers, this will completely flush the memory where the firmware is run from which should help get rid whatever remains of the attempts to access your router.

3. If no update has been released for your router in the past year to two year time frame, looking at buying a new router today

Using a $50 router from your local office supply store or electronics retailer won't do that much to help you. At that point, it is just a matter of time before your router will make someone's list for getting hacked. Nothing is guaranteed to keep someone from getting in. The trick is to make it difficult enough to encourage them to move on to easier lower hanging fruit.

The cheapest routers available are known as stateful routers. This is a fancy way of saying that they will only allow traffic in that can be matched to an outbound request from a pc or device on your site of the router/firewall. In an ideal world, this should prevent an incoming attack but it doesn't always work that way.

The next generation of firewalls/routers take protecting your home network to the next level. They look at traffic from both directions and look a little deeper into the communications that are going on. You may hear the term "deep packet inspection". This means that the router/firewall is looking past the ip addresses in each packet and more into what is going on. Another term to look for is "heuristic analysis". In this case, the traffic is looked at for signs of what isn't normal or expected. Heuristic analysis doesn't have to look at potentially personal information to see a problem the way deep packet inspection might have to.

What should happen is that anytime something bad is found, that information (minus anything that identifies you, your router, etc.) should go back to the manufacturer for review. If a problem is found, an update to your device should be pushed out within an hour or so. This type of functionality has been available in the commercial world for several years and has finally made it to the consumer market.

What this means is that your next firewall/router should be more than a set and forget situation where you buy the new device, unbox it, and plug it in. Along with this will come a recurring cost, either on a monthly or yearly basis for the update service. Think of it just like another utility bill. Would you rather pay a little bit now on a recurring basis to avoid problem or even more money to deal with the mess after a problem has occurred.

As we all increase the number of devices in our homes, we are creating our own "Internet of Things" network. Because those devices might be compromised at some point, this is another reason to think about things a little differently. Ideally IoT devices should be on their own separate wireless or wired network. In the event that something happens that cant be easily fixed, you can quickly cut off all access to the internet and your home network for those devices.

This another reason to look at the "Next Generation" firewalls/routers that I mentioned earlier. Part of their "magic sauce" is they profile or establish a baseline for what is normal for a particular device on your network. When communications deviate from that level of normalcy, that is when the device steps into action and shuts off access for that device to the Internet.

Anytime you have this type of protection in place, there is bound to be some level of "false positives". There are events that are identified as a problem but really aren't. Which would you rather have - some false positives or not knowing how safe your network really is. I go for the false positives options versus not knowing anything.

Ethernet over Powerline

There will be times when setting up your Smarthome Network that you will need to choose between using wireless or a wired connection. Depending on the age of the home you are implementing Smarthome tech in, getting a good wireless signal from where your wireless router is to where the device that needs to use it is may be a real challenge. Running a ethernet cable from where your internet connection is to where you need it to be can be both a lot of time and frustration. Cutting into walls and drilling in the floor to get the cable to where it needs to be will be messy and potentially expensive.

The device I am talking about are now in their second generation and have improved a lot over what they were initially. Basically, they convert the ethernet connection you need to running over the AC electrical lines in your home. Don't worry about damaging the equipment you will plug into the EoP (Ethernet over Powerline) devices. I have yet to see or hear of any damage from doing so.

There are two main powers of EoP adapters you can choose from, those with an outlet on the front of the device and those without. The cost difference between the two types will be a few dollars at best based on the ones that I have purchased for my one use. An important thing to remember is that for these devices to work, they will need to plug in directly to the wall outlet. If you plug them into any type of surge protection outlet strip, they either wont work at all or will be very sporadic/intermittent. The same concern applies to when you plug them into either GFI or similar outlets. The circuitry inside the the GFI outlets will cause problems for the EoP devices.

To get them working couldn't be easier. You will need to pair the devices together. I usually take an extension cord with multiple outlets. Plug both devices in the extension cord, wait for them to power up and then start the pairing process. Getting that done beforehand will make things a bit easier when you put them into production.

You will see a variety of speeds available. Keep in mind the speeds quoted should be thought of as theoretical. You might see those speeds or you may get close to the speed. Look at the speed of your internet connection as a part of the selection process. If you only have 200 MB internet connection, getting EoP adapters capable of Gigabit speeds will be money spent that you probably wont see the full benefit from. If most of your use will be device to device and not as much over the Internet connection, then Gigabit rated EoP adapters may make more sense.

One of the first things I would recommend on this type of equipment is that after you have them installed and configured, check the manufacturers website for the latest firmware updates. My experience is that most of the time when I purchase something, it rarely comes with the latest firmware installed. As a part of this download process, get the latest management app for the devices. When you are setting up the initial pairing of the devices, the encryption key used may not be the strongest. Using the management app should give you the ability to use a much stronger key and therefore increase the difficulty for someone to be able to crack the connection between the devices.

If you end up needing a wireless signal where you are placing the EoP devices, there are some of them that also offer a wireless AP functionality. The main concern here is that you now have yet another wireless access point to configure. Especially on the EoP devices that have wireless functionality, it is even more important to make sure the latest firmware is installed.

Video

https://youtu.be/5aYU36B8D7g

TRENDnet Powerline 500 AV Nano Adapter Kit

http://ronaldnutter.com/TP500AV

TP-Link AV600 Powerline Adapter Starter Kit
http://ronaldnutter.com/tpav600

Recovering from a power outage

The troubleshooting suggestions that I have here are meant as a starting point. Once you get comfortable with everything that is or will make up your IoT network, some of the troubleshooting steps or methods may be a little different. Once you have a baseline of how your particular combination of IoT devices work, the easier you will find it to troubleshoot when necessary.

Things to do before a power loss

1. Backup of the configurations of the IoT devices that have that function. For those that don't have this option, make sure you have it documented how they were configured, device id's or activation codes, etc., just in case you have to do a factory reset.

2. Have a test plan ready

This isn't has hard as it may sound. Looked at the devices you use the most daily and how you use them (turning on a light, turning on a media receiver and selecting a favorite movie, etc.). Write a step by step of what to do and list the expected result. Sometimes, even a small different in the result could indicate a problem that needs further attention.

After Power is restored

1. Give everything time to startup and connect

When power is restored, your first inclination is the go back to your normal routine. After just having been through a power outage that lasted over 30 hours, I learned some interesting lessons. The first one is to allow all my IoT devices about 15 to 20 minutes to settle down before trying to use them. Device such as my Roku and Fire TV were among the first items to start working again, even before the Google Fiber Network box showed that it was finished going through the startup process.

Some devices, such as anything that had to talk through an intermediate hub such as Samsung Smartthings took a lot longer to start working. This is not to pick on Samsung but this just happen to be the device that I had the most problems with. It took nearly a day to get this one working again. The hub indicated that it had both power and a good network connection but it would now show as online in the Smarthings smartphone app. Unplugging and replugging the network cable into the Google Fiber network box didn't resolve the problem. Moved the network cable to a downstream network switch and it started working.

2. Single device problem

You may need to try testing a device more than one. Especially after an extended power outage, it may take at least one use of the device in question before things start to work as expected. if the device still is having a problem, power cycle the device and give it a few minutes to finish the startup process. Go into the configuration page for the device and see if there is anything that doesn't look right. If nothing looks wrong, trying changing the setting of one option, save the config and then change the option back to the original setting and save again. If there was a minor corruption of the device's configuration, this should help resolve it. If the problem persists, it may be necessary to do a factory reset of the device and reconfigure it. If a factory reset doesn't fix the problem, you can try reaching out to the vendor's technical support to see if there are any other steps to try. It may be that the device failed when the power was restored and needs to be replaced.

3. Multidevice problem

In a multi-device problem, if there is a IoT hub (I.e. Z-Wave or Zigbee type devices) is involve that the devices have to connect or associated to in order to work, that could be the easiest place to start by restarting the device. If the hub still isnt showing up in its management app, try unplugging and replugging work cable at both ends of the cable. In one case, I had to go as far as moving the IoT hub to a different port on the switch before things started working correctly. Even if some devices are working, it still might be a IoT hub issue. If restarting the hub doesn't resolve the problem. Try restarting one or more of the devices that connect to the IoT hub in question. If you don't see the devices re-establishing, have the IoT hub "forget" one of the devices. Then do a factory reset of the device and configure it to talk to the IoT hub.

Options for avoiding potential IoT device damage from a power loss

1. Turning off all breakers that have IoT devices present while power is down

a. Waiting until power has returned for about 5 to 10 minutes to see if it will stay up for a while before you start bringing the IOT device back up.

2. Using Surge Protectors

a. When possible, use a surge protector to help prevent a power surge when power is first restored from damaging the device.

b. Surge protectors don't last forever. Over time, their effectiveness can gradually weaken each time they have block a power surge from making it through to the end devices. You should consider replacing surge protectors at a 2 to 3 year interval. Waiting until the light that indicates the protector is working to go out might run the risk of a power surge of damaging one or more devices.

3. Using a UPS to protected streaming media receivers, cable modem and IoT hubs

UPS's are a good thing. Keep in mind that they wont' keep what is plugged into them up forever. Having more battery in the UPS than you have devices plugged in will govern how long the devices will stay up before shutting down if power hasn't been restored.

There are two kinds of UPS's, those that only protect from power failure and those that protect from both power failure and power surges. Go for the latter whenever possible. As a general rule, the UPS should be plugged directly into a wall outlet. I have seen the electronics in a UPS get damaged when plugged into a surge protection outlet strip because of what the surge protection strip does in filtering the power. Not as likely to happen but there is a chance.

Terminology

This is an area that I have waited to cover until now. I felt it was more important to get you familiar with the different ways of controlling lights or other devices. For those that would like a little more information on the technology behind what you have been doing, that is what we will start to cover in this chapter.

Zigbee is one of the standards available in the area of Home Automation that allows device from more than one vendor to talk to each other. Since Zigbee is low power, it offers the potential of running on a single battery or set of batteries for several years. If the Zigbee device uses the Green Power Feature, you won't need batteries for the device to function. Zigbee uses the 4Ghz ISM (industrial, scientific, and medical radio band) so it can operate license free around the world. For more information about this, please go to http://www.zigbee.org.

Z-Wave is a mesh based technology. One Z-Wave device has a range of 120 feet between one device and another. With the ability of Z-Wave device to function as a repeater, it is possible to achieve an effective range of 600 feet. Each Z-Wave network can support up to 232 devices. Z-Wave using a 1Ghz frequency range.

802.11ac refers to a wireless networking standard that if you don't already a router on your home network that supports this, you should get one soon. Using this wireless standard in conjunction with a 5Ghz wifi band will go a long way into getting you the best performance on streaming media receivers.

IOT which is short for internet of things. This typically refers to devices that may not have a keyboard, display or mouse but need access to some type of network. As you start expanding on the smart home devices, you are building your own Internet of Things.

POE stands for Power over Ethernet. This is a way of power a device and supplying network connectivity at the same time. There are two standards for doing this - the way that Cisco initially implemented and a standards based implementation known as 802.3af.

Where do I get started ?

There are several ways to get started. I would first suggest deciding on what you would like to look at first. For some, it could be selecting a streaming media receiver. That would help increase your learning by being able to watch YouTube videos on your television which have the potential of helping you learn even more about what you can do with your potential new smart home. You should have already noticed that I have several videos linked to in this book that go into more detail or show you better with visuals than with the written word. It is entirely possible that you may start with one streaming media receiver and then add a different one from another manufacturer due to one or more features that one vendor has that another doesn't.

The next area to look at is what around your home that you would like to increase the utility of. How many times have you left the house only to return late and with no lights left on to make someone think that there is someone in the house. With smart home technology, you have different ways this can be handled. Using a smart bulb or outlet connected to a light, you can remotely turn on or off a bulb without having to be a home. Depending on the system you implement, it may be possible to schedule the light to turn on or off at a specific time or automatically adjust that sequence based on the time the sun sets or rises.

Another area to look at is adding voice control. You shouldn't have to use a remote control app from one or more vendors to control each of the devices. Having voice control is an option that is not only nice to have but could make smart home technology the opportunity to add quality of life to someone that is physically challenged and not able to turn lights on and off around the house as easily as some of us do. That is just one example of what this type of technology has the ability to do.

You can reach out to me via my YouTube channel with questions that you have. If I have a video that answers your question, I will be happy to give you that length. If I don't have a video I will look at seeing what I can do to create one or give you information that will help you in what you are trying to do.

Please check out my other Books
and add them to your library today !
Please visit http://www.ronaldnutter.com for more info

This book will help walk you through the process of protecting your online accounts from hackers and those who like to use your accounts for something other than what you intended. Most website and social media accounts can be protected in as little as 3 steps. Advanced methods of protecting your accounts such technologies such SMS and two factor authentication.

http://www.ronnutter.com/kpyop

Are You Disaster Ready? - Communications

http://www.ronnutter.com/ppv1c

You can't always depend on a cell phone during an emergency or disaster. This book covers a variety of options such as Satellite, Amateur Radio, MURs, GMRS and several other options. One thing you need to read about is how to use your cell phone to establish your own private texting network !

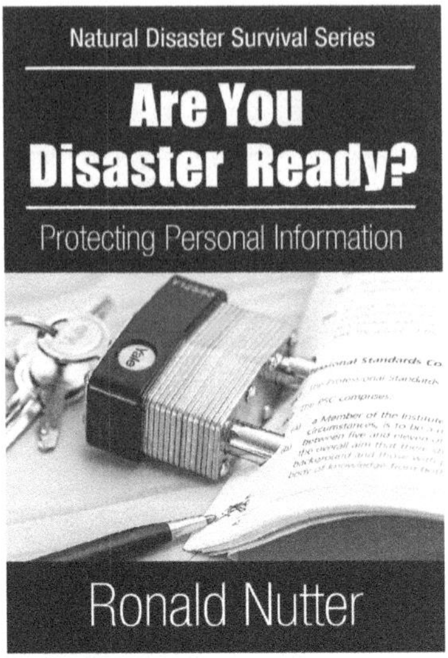

Are You Disaster Ready? - Protecting Your Personal Information

http://www.ronnutter.com/ppv2p

This book covers steps you will need to take to protect your personal information during a disaster. This is a prime time for identity thieves to strike. This will also be the same time when you will have the least time and resources to deal with the situation.

Are You Disaster Ready? - Food Preparation

http://www.ronnutter.com/ppv3f

Depending on the length of a disaster or emergency situation might mean that the grocery store(s) near you will be low or out of food. Know how to do some basic food preparation steps or even growing some of your own foot before a problem occurs will help you be that much more self sufficient for the duration of the disaster/emergency and as a matter of general principle.

Depending on the length of a disaster or emergency situation might mean that the grocery store(s) near you will be low or out of food. Know how to do some basic food preparation steps or even growing some of your own foot before a problem occurs will help you be that much more self sufficient for the duration of the disaster/emergency and as a matter of general principle.

This book is targeted at the new VMware Admin or someone who has been out of the field for a while and needs to come up to speed fast. Think of this as an electronic notebook with things you won't learn in a VMware class. If you are looking for a job in this area, some of what you find here might just make you stand out from the others interviewing for the position.

http://www.ronnutter.com/vmna1

If you have dealt with any type of technology, you will know there are times when using the GUI won't work. This book gets you familiar with the different ways of accessing either a VMware host, vCenter server or VDP appliance. Knowing how to navigate around the Command Line Interface in the VMware product family is a handy skill to have.

At some point in your VMware career, you will need to do at least on upgrade. While no two upgrades may ever be the same, having a rough blueprint will make the process a little easier. This is another area where doing a dry run in your lab will help give you an idea of what to expect in the real world. Running into problems in the lab is easier to deal with than in a production environment.

http://www.ronnutter.com/vmna3

In volume 4 in the VMware - A Guide for New Admins series, I put together a timeline for those with little to no background in VMware to help them get started on the path on learning VMware and a new job skill. This book will be updated as I release additional books in the series to continue the buildout of the path for increasing your VMware skills.

http://www.ronnutter.com/vmna4

This book is targeted at the new VMware Admin or someone who has been out of the field for a while and needs to come up to speed fast. Volume 5 deals specifically the different things you need to think about in backup up your growing VM farm. This is another area where you will need to be well versed in.

http://www.ronnutter.com/vmna5

It is estimated that somewhere between 1 and 5 to upwards of 50 million people suffer with Tinnitus, otherwise known as ringing in the ears. In the past, there wasn't much that could be done to treat Tinnitus as there currently is no cure for the problem. That all changed in early 2014. This book is a diary of my journey since being diagnosed with Tinnitus and what I have learned along the way.

http://www.ronnutter.com/tims

Android isn't just an operating system for your phone or tablet. There are a variety fo apps that can turn your device into a ad-hoc network testing tool so you don't have to carry a war chest of testing equipment and tools with you. With the tools listed here, you can be ready to troubleshoot at a moments notice !

http://www.ronnutter.com/asne

Made in United States
Cleveland, OH
05 November 2026

47664000R00142